Royal Scottish Museum Studies

Joseph Black 1728-1799
A Commemorative Symposium

Papers presented at a Symposium
held in the Royal Scottish Museum
on 4 November 1978
in association with
The Scottish Society of the History of Medicine,
together with a Survey of Manuscript Notes of
Joseph Black's Lectures on Chemistry

Edited by
A. D. C. Simpson

The Royal Scottish Museum
Edinburgh 1982

ISBN 0 900733 25 X

Contents

Foreword

Joseph Black is rightly regarded as an influential figure in 18th century science. His development of the concepts of latent and specific heats and his role in establishing pneumatic chemistry have reserved for him a special place in the history of chemistry. As an exponent of the new philosophical chemistry Black was a significant force in the Scottish Enlightenment, playing a major part also in the exceptional success of the Edinburgh Medical School.

The 250th anniversary of Black's birth fell in 1978 and it was felt that this provided a suitable occasion for a commemorative symposium at which some aspects of Black's work could be examined. It was appropriate that such an event should have been held at the Royal Scottish Museum since Black's surviving apparatus is preserved here, and we were delighted to have been associated on this occasion with the Scottish Society of the History of Medicine.

Norman Tebble, D.Sc.
Director

Figure 1. Joseph Black: portrait in chalks by an unknown artist, c1790.
(Reproduced by permission of the Scottish National Portrait Gallery)

Preface

The papers printed here were presented at a symposium held in the Royal Scottish Museum on November 4th, 1978, to mark the 250th anniversary of the birth of the chemist Joseph Black. This was held as a joint meeting with the Scottish Society of the History of Medicine, and with the support of the Royal College of Physicians of Edinburgh and the Royal Society of Edinburgh. The meeting was organised by my former colleague A.Q. Morton, who is now at the Science Museum, London.

The programme was not intended to present a balanced view of Black, but rather to highlight certain facets within the confines of a one-day meeting. R.G.W. Anderson and C.J. Lawrence have provided an outline biography and an assessment of the philosophical context of Black's work. Henry Guerlac has examined the evolution of Black's ideas on change of state and specific heat, and Peter Swinbank provides here a background to the teaching of science in Glasgow during Black's period at the University. Black's success as a prominent physician, an aspect of his life that has received scant attention, is discussed by Andrew Doig, whereas W.P. Doyle has been concerned with Black's accommodation of Lavoisier's chemistry in his teaching. Finally, J.R.R. Christie has discussed the extent to which the received picture of Black has been coloured by the attitude of his editor and disciple John Robison.

In addition to these papers, the volume includes a detailed survey by William A. Cole of extant sets of manuscript student notes of Black's chemistry lectures, intended as a tool for the study of Black's adoption of Lavoisier's doctrine.

The Royal Scottish Museum's 'Playfair Collection' of early chemical apparatus, which includes the surviving teaching apparatus from Joseph Black's Edinburgh laboratory, has been described in detail by R.G.W. Anderson in a companion to this volume. The publication of Dr Anderson's *The Playfair Collection and the Teaching of Chemistry at the University of Edinburgh 1713-1858* (Royal Scottish Museum, 1978) was timed to coincide with the Joseph Black symposium.

A.D.C. Simpson
Department of Technology

Contributors

Dr. R.G.W. Anderson (speaker & chairman)	Deputy Keeper Wellcome Museum for the History of Medicine The Science Museum London (now Keeper, Department of Chemistry, The Science Museum)
J.R.R. Christie	Lecturer Division of History and Philosophy of Science Department of Philosophy University of Leeds
William A. Cole	815 El Medio Avenue Pacific Palisades California
Dr. Andrew Doig	Physician Royal Infirmary of Edinburgh
Dr. W.P. Doyle	Senior Lecturer Department of Chemistry University of Edinburgh
Professor Eric G. Forbes (chairman)	Professor of the History of Science Department of History University of Edinburgh (now of the History of Medicine and Science Unit, University of Edinburgh)
Professor Henry Guerlac	Professor of the History of Science Department of History Cornell University Ithaca New York
Dr. Christopher Lawrence	Medical Historian to the Wellcome Museum The Science Museum London
A.Q. Morton (chairman & organiser)	Assistant Keeper Department of Technology Royal Scottish Museum Edinburgh (now at the Department of Physics, The Science Museum, London)
Dr. Peter Swinbank	Senior Lecturer Department of the History of Science University of Glasgow
Dr. Haldane P. Tait (chairman)	President Scottish Society of the History of Medicine Edinburgh

Joseph Black: The Natural Philosophical Background
Christopher Lawrence

In the last quarter of the 18th century the medical
school of the University of Edinburgh was rocked by the
Brunnonian controversy. In this dispute, ostensibly about
medical theory, the former Edinburgh student, flamboyant
iconoclast, Jacobite and drunkard, John Brown deployed
his forces against those of the established medical faculty.
A fictional episode in this considerable event was recorded
by a follower of Brown in an epic poem, *The Brunnoniad,*
in which the antagonists, thinly disguised as the ancients,
confront each other in an Edinburgh tavern. In his laboured
couplets the author portrays the august professors of the
medical faculty in colours somewhat less flattering than
those in which they are usually seen.

William Cullen, undoubtedly the most renowned teacher
of medicine in the British Isles he depicts as Nestor. The
butt of the author's criticism is the vast hiatus between
Cullen's elaborate theorizing and his dangerous, incompe-
tent medical practice.
> "Nestor, who now that sable garment wore,
> Which many a grave professor deckt of yore,
> White as the milky dove, or Boreal snows,
> His ample wig around his shoulders flows,
> And seventy rolling years in vain control
> The flights eccentric of his daring soul.
> From noise secluded in his airy cell,
> Where proud Philosophy delights to dwell,
> Still as in youth intent on bold designs,
> Line into Line, and page to page he joins;
> In painful study yet exhausts his skill
> To form a bolus and to mould a pill;"[1]

He has a similar comment to make on the professor of
physiology, James Gregory:
> "Tall as the Highland fir; with thund'ring stride
> Machaon enter'd: o'er his learned head
> Few scatter'd hairs in rude disorder spread,
> By study thinn'd; all ornament he scorn'd,
> Nor with broad wig his naked scalp adorn'd.
> In theory's mazy paths his mighty mind
> Excell'd, and left his fellows far behind.
> In practice great, his quick unerring hand
> Sent many a wretch to Pluto's dreary strand."[2]

And so he goes on being none too complimentary about the
other Edinburgh worthies, Francis Home and Andrew
Duncan. One man, however, escapes this vilification,
Joseph Black:
> "With pond'rous staff, next Paracelsus came,
> In chemic arts who boasts unequall'd fame.
> His modest aspect cast upon the ground,

Instant he past, as fixt in thought profound . . .

> None better knew to raise the glowing fire,
> And bid the pungent volatiles aspire.
> Sound his opinion, and his judgment rare,
> Of salts and earths, of sulphur and of air."[3]

The Brunnoniad however, demands attention beyond its
merits as a piece of doggerel, for it contains almost inci-
dentally a precise and illuminating characterisation of two
of Edinburgh's most esteemed professors. William Cullen's
"eccentric" theorizing I will suggest was something more
than an idiosyncrasy. Rather it was deeply rooted in the
Scottish philosophical context. Black's "modest aspect"
on the other hand was that very feature of his enigmatic
personality that endeared him to his contemporaries for it
sheltered them from his opinions on the broader issues of
religion or philosophy. Black rarely exposed his deepest
thoughts on nature to public display.[4] It may however be
possible to reclaim Black's philosophical outlook by
inference if not by direct access. In the first part of this
paper I shall describe the major changes in British natural
philosophy that Black would have witnessed and in the
second part I shall try to show how these changes took a
specific form in the Scottish context. To do this I shall
examine the views of Black's closest friends. Black's own
opinions I shall leave, as he did himself, as a source of
conjecture.

Joseph Black was born in 1728, one year after Newton's
death. Black himself died in 1799, the year of John Dalton's
first publications which were to lead him to his atomic
theory. These events can be taken to signify that it was
during the 18th century that chemistry enthralled itself to,
and finally emancipated itself from, mathematical physics.[5]
This was but one of the many significant intellectual
changes that coincide with the period of Black's life. When
Black was a five year old child, Stephen Hales, the Vicar of
Teddington, published his *Haemostaticks,* an experimental
investigation into the dynamics of the blood and vascular
system, by the time of Black's death Charles Bell was well
advanced into his project of producing a new model of the
nervous system. Two important intellectual revolutions
were signalled by these events. First a move from a
mechanist to a vitalist approach to physiological problems,
and second a displacement of the centre of interest from
the vascular to the nervous system. This latter shift was
not unrelated to another of a far more general kind. The
nervous system in the 18th century became the repository
of the property of sensibility and it was this same century

after all, which witnessed a change in the concept of the basis of human social activity. Reason gradually gave way to the sentiments, feeling triumphed over rationalism.

Black's life then cannot fail to be of interest, not least because his most creative work was done at the very period when the most spectacular intellectual changes of the century were occurring. Indeed, many of the men who were to be the most important theoreticians of the new directions in 18th century thought were Black's country-men, his colleagues and, in a few cases, his close friends. During Black's early years David Hume produced his sceptical philosophy, Adam Smith created his theory of the moral sentiments and Robert Whytt did his great work on the nervous system.

Over the whole of 18th century thought, long after his death, Sir Issac Newton cast his shadow. His *Principia Mathematica* had appeared in 1687, the culmination of a 200 year revolt against the Aristotelian cosmology. It meant, as Dijksterhuis put it, "the introduction of a description of nature with the aid of the mathematical concepts of classical mechanics."[6] The fruitfulness of this programme for subsequent mechanics and astronomy is obvious enough, yet it was not an explanatory theory from which chemistry and physiology could draw without recourse, at a more intermediate level, to theories of matter, power and the causes of change. But here, of course, there was no inflexible and agreed mathematical basis, but only the extremely diverse speculations that filled Newton's later works especially the *Opticks*. It is in these specula-tions that recent historians have found much of the history of 18th century natural philosophy.[7]

Newton's own theory of matter is summed up in the well-known Query 31 of the 1717 *Opticks*:
"all things being considered it seems probable to me that God in the beginning formed matter in solid massy, hard, impenetrable, movable particles of such sizes and figures and with such other properties and in such proportions to space as most conduced to the end for which he formed them."[8]
What was so different about Newton's corpuscular philo-sopy compared with that of his forerunners was that his posited particles possessed not only the properties of extension and motion, but were to be regarded as endued with powers, propagated through space, and serving to attract and repel other bodies as a function of their distance. Now in speaking of such powers, of which gravity was the model, Newton insisted he was only speaking mathematically since the notion of matter possessing forces acting at a distance smacked of the occult qualities of scholasticism which had, for the mechanical philosophers of the 17th century, become an anathema.

It was on the attractive and repulsive properties of such particles that Newton conceived the operations of chemistry to depend, and, conversely, it was through chemistry Newton hoped that some measure of these forces might be attained. "There are therefore" he said:
"agents in nature able to make the particles of bodies stick together by very strong attractions and it is the business of experimental philosophy to find them out."[9] Chemistry, in other words, was to be an appendage of physics. Newton's particles with their associated forces and his vision of a quantified chemistry were to prove re-currently appealing to chemists throughout the century. They were also to prove so impossibly complex and far from experience as to be virtually useless as the basis of a chemical theory. Given the figurative shadow of the great man's ideas and his literal presence at the Royal Society it is not surprising that the first generation of Newtonians, John Keill, John Freind, Francis Hauksbee etc. should experiment and speculate in physiology and chemistry in this reductionist manner concentrating on the short range forces between particles which, in the event, it proved impossible to quantify.

Newtonianism however was not merely a British phenomenon, for it found continental disciples too, a fact of some importance for Scotland. Though in France it generally encountered a hostile reception, in Holland Newtonianism was warmly received by Herman Boerhaave whose name was synonymous with the Leyden School of Medicine. Boerhaave's chemistry, which was ultimately subservient to his medicine, drew on a variety of traditions besides Newtonianism but it did endorse Newton's view of chemistry as arising from interparticulate forces. Boerhaave's chemistry was to be the first chemistry taught at the new Edinburgh school, and therefore an important source of Scottish Newtonianism.

There is one further significant episode to be recorded of the endeavours of the first generation of Newtonians. In 1727 the Vicar of Teddington, Stephen Hales, published his *Vegetable Staticks*. From a series of experiments Hales had discovered that atmospheric air — a single substance — could become fixed in solid bodies and played a significant part in chemical processes. This air, when released, Hales regarded as being composed of mutually repelling particles. Hales' work thus inaugurates the era of gas chemistry whose next significant practitioner was to be Joseph Black followed by Henry Cavendish, Joseph Priestley and Lavoisier.

Chemistry in early 18th century Britain served two masters: Newtonian physics in theory, and medicine in practice. Gradually, during the century it emancipated it-self from both of these. On the social level chemistry began to emerge as a profession aligning itself to incipient industrialisation. On a theoretical level British chemists like Black and Cullen, though they did not abandon their Newtonian ideals, moved gradually away from reduction-ism and in practice worked at the phenomenal level of chemical properties.

These significant changes in the practice of chemistry go hand in hand with a similar move in physiology. Here the change was from a dualist physiology based on the model of the body as a machine to a monist view based on the phenomenal properties of irritability and sensi-bility conceived of as residing in the muscles and nervous system. Such striking conceptual changes were principally formulated by Black's countrymen, Robert Whytt and William Cullen. These changes, which might be seen as part of a general move away from reductionism towards the phenomenal, mechanism to materialism, as Robert Schofield has called it, were associated with an important theoretical reorientation.[10] A shift not unconnected with Black's work on heat.

Newton had postulated that the forces between particles were purely mathematical descriptions, however he, and still less the 18th century, could not resist speculating as to what the nature of these forces might actually be. In the second edition of the *Principia* of 1713 Newton added a general scholium to Book III which went as follows:

"And now we might add something concerning a certain most subtle Spirit, which pervades and lies hid in all gross bodies; by the force and action of which Spirit, the particles of bodies mutually attract . . . and electric bodies operate . . . as well repelling as attracting the neighbouring corpuscles, and light is emitted, reflected, refracted, inflected, and heats bodies . . ."

This was the famous Newtonian aether on which Newton elaborated in later editions of his *Opticks*. Exactly what Newton conceived the aether to be is difficult to fathom. It was a subtle elastic spirit composed of minute particles perhaps mutually repelling pervading all matter and being the immediate cause of gravity, magnetism, electrical phenomena and so forth. It seemed to partake of both the material and immaterial realms.

To the first generation of Newtonians, the protagonists of a dynamic corpuscular philosophy, the aether was possibly a source of some embarrassment for it is hardly mentioned. By mid century, however, the aether had begun, like the subtle fluid it was, to insinuate itself into natural philosophy.

The first important publication on this mysterious substance was Bryan Robinson's *Dissertation on the aether of Sir Isaac Newton* of 1743. From here the aether entered 18th century science in many forms; as the principle of heat, electricity, magnetism and as the active agent in the nervous system. Another important source for the theoreticians of the aether was Boerhaave's doctrine of fire. Fire for Boerhaave was an imponderable fluid dispersed throughout matter, it was he said "the great changer of all things in the universe while itself remaining unchanged."[11] By mid century the idea was widespread that fire, aether and electricity were all modifications of the same substance.

This changing theoretical basis of mid 18th century science was more than just a change in the explanatory model since it entailed in some cases a radically different conception of the autonomy of nature.[12] For Newton, all the laws of nature were imposed by God, and active principles in nature were manifestations of the divine agency. All activity, in other words, was God's activity. By mid century, however, many thinkers had come to regard nature as self-sustaining and self-sufficient. One of the most important theoreticians of this concept was Black's friend,

James Hutton. In Hutton's theory the source of all activity in the universe is an aethereal substance emanating from the sun which in different modifications appears as electricity, heat and light.[13] Aether too, it should be added, was also central to the thought of Black's mentor William Cullen.

Turning to the natural philosophical background of Black's work in its Scottish context it is possible to see these same general changes occurring but in rather specific ways depending on the distinctively Scottish setting. In intellectual terms the important currents in Scottish thought that are seen in Black's ideas are nearly all associated with men who at some time were professors at the Edinburgh medical school. The medical school was founded in 1726. The first five professors in the faculty had all trained at Leyden under Herman Boerhaave. But it was not only the Leyden curriculum they established in Scotland but, almost in its entirety, Boerhaave's medical and natural philosophical system. Adoption of this system meant commitment to several important doctrines. First, to a corpuscular theory based on short range forces, and a corresponding rejection of aether theory. Alexander Monro the first professor of anatomy dismissed Robison's work on aether for taking as he says "Sir Isaac's queries as axioms".[14] Second, the adoption of a strictly dualist model in physiology. The body is an hydraulic machine with an immortal soul related to it in some wholly unknown way. Life is no more than the incessant motion of the body's particles particularly those of the blood. Finally, such natural philosophy is seen as the cornerstone of religion in so far as science is a demonstration of final causes. This conception is found most clearly described in the works of the other great Newtonian of this generation, the professor of mathematics Colin Maclaurin. Maclaurin says:

"Natural philosophy is subservient to purposes of a higher kind and is chiefly to be valued as it lays a sure foundation for natural religion and moral philosophy; by leading us in a satisfactory manner to the knowledge of the author and Governor of the universe."[15]

Methodologically this generation was characterised by a commitment to an unsophisticated Baconian programme of observation and experience, scorning any apparently speculative systems. This approach was enshrined in the famous *Medical essays and observations,* the product of the first generation of professors, a work which looked unfavourably on articles deficient in "facts".[16] The first professor of chemistry in the new school was Andrew Plummer, most famous as the formulator of Plummer's Pill, a compound of antimony and mercury. Plummer's chemistry enshrined the reductionism of Boerhaave and Newton's early English disciples. Plummer's chemistry lectures, however, were a fairly straight forward account of chemical facts and processes. Donovan reports that Cullen found Plummer's lectures 'old fashioned' and 'unsuggestive'.[17]

Although there is much continuity in the history of the Edinburgh medical school, there are also some striking differences in doctrine differentiating the teachers of the second generation from their predecessors. The leading theoretician of this new generation was undoubtedly William Cullen. Cullen had come to Edinburgh from Glasgow in 1755 to be professor of chemistry. Cullen was proud to proclaim that he had rejected a great deal of the Boerhaavian system of medicine.[18] This indeed was the case in a very general sense, though less so in terms of detail. Cullen's physiology was based on the nervous system whose most important property was sensibility. It was through this property and its associated response, irritability, that all the involuntary activities of the organism were performed. But sensibility was also the capacity to perceive and to have ideas and therefore from John Locke's philosophy the basis of thought. In other words, Cullen had begun the reintegration of mind and body which had remained severed in orthodox medicine since Descartes' infamous 'cogito ergo sum'. Monism, in other words, was replacing dualism. A major part of Cullen's theorising on the nervous system was his doctrine of the aether, that fluid which when in an excited state constituted life itself by giving rise to sensibility and irritability.[19] The aether too, figured substantially in his theorising on the nature of fire and light.

This total reorientation of Edinburgh medicine by mid century was not merely accidental but can be linked fairly specifically to some major changes in Scottish philosophy and indeed Scottish society generally. In 1739 David Hume had published his *A Treatise on human nature*. An important constituent of this work was Hume's analysis of causation. Basically Hume contended that when one set of impressions is constantly followed by those of another set we tend to form a belief in a necessary connection between them, attributing causality to the former and effect to the latter. Hume points out, however, that all we can actually say is that these events are constantly conjoined in our minds, that is, it is a psychological fact and not a logical necessity. Such a conclusion of course had radical implications for scientific knowledge, which is distinctly circumscribed in what it can achieve. Hume's conclusions were fairly rapidly incorporated into all orthodox Scottish philosophy, but with two radically different results.[20]

Hume's sceptical philosophy, which clearly had implications about the existence of God had drawn a fairly rapid reply from a minister, Thomas Reid, whose answer eventually became the basis of what was to be known as the Scottish philosophy of common sense. In essence this philosophy asserts that there are certain elementary truths, such as the existence of matter and God, which we know without experience and cannot therefore be the subject of sceptical philosophy and thus of doubt. Reid did, however, adopt Hume's conclusions about the nature of scientific knowledge the aim of which he saw as the description of phenomenal events and possibly later, by induction, arriving at some rather more general laws. Such a view of science was incorporated into a thoroughgoing defence of Christianity for it was seen as revealing those general laws of nature which were God's handiwork. It was a tradition which was vigorously anti-conjectural eschewing any hypothesis about non-observable events. This philosophy, which was to become the dominant one in Scotland, was

represented at the medical school by Thomas Reid's cousin, the professor of physiology and medicine, John Gregory. As part of its anti-conjectural campaign the aether came in for particularly violent attack. It was an hypothesis and ergo unacceptable, but one might also note it was, in the more suspect philosophies such as that of James Hutton or Joseph Priestley, the principle of activity in the universe and therefore a usurper in the role of God. Thus the aether served ideological as well as scientific ends. It might be pointed out that there is a rather strong continuity with this mid century common sense formulation and the doctrines of the earlier Boerhaavians.

The other philosophical tradition stemming from Hume remained centred around David Hume himself and had few explicit practitioners. This tradition adopted Hume's conclusions about the nature of scientific activity, yet at the same time recognised the necessity for speculation and conjecture. For Hume such speculations were necessary in order to provide physical continuity in nature to make the action of a cause comprehensible. Another formulation of this conjectural alternative is found in Adam Smith's *History of Astronomy* in which Smith finds speculation to be a psychological necessity since by making nature coherent it provides mental composure. In terms of practising scientists the man most closely aligned to Smith and Hume in social and intellectual terms was William Cullen. Cullen speculated endlessly on the action of aether in nature and drew angry criticism from Gregory for his pains. But the reason for this seems to be not only that speculation was an anathema to Gregory, but that in Cullen's philosophy the aether took the place of God. Gregory had reason for complaint.

There are here, I think, clear implications for an understanding of Black's natural philosophy in the widest sense of that term. For it is certain his personal friendships and intellectual debts aligned him with Hume, Smith, Cullen and Hutton. However Black's own views on God, aether, and the activity of nature remain something of an enigma. But it was that very reticence in his character that protected him from criticism. For the Brunnonian rhymster there was no one, not even John Brown or Bruno as he called him, as erudite as Black. But in saying so he also hints at Black's avoidance of polemics that makes the discovery of his real views so perplexing to the historian.

"Disputes he shunn'd, nor car'd for noisy fame;
And peace forever was his darling aim.
Calm in the hall he took his peaceful seat,
In philosophic lore not Bruno's self more great."[21]

NOTES AND REFERENCES
1. Anon. *The Brunoniad* (London 1789) 6. The author was probably William Margetson Heald.

2. *Ibid.* 23.

3. *Ibid.* 22.

4. Nothing, for example, is known of his religious persuasion, if any.

5. Arnold Thackray *Atoms & Powers* (London 1970).

6. E.J. Dijksterhuis *The Mechanization of the World Picture* (Oxford 1969) 501.

7. The literature is too large to cite but a survey can be found in Steven Shapin "The Social Uses of Science 1600-1800' in R.S. Porter and G.S. Rousseau (eds.) *The Ferment of Knowledge* (Cambridge 1980).

8. Isaac Newton *Opticks* (London 1717).

9. *Ibid.* 369.

10. Robert E. Schofield *Mechanism & Materialism* (Princeton 1970).

11. Hermann Boerhaave (trans. Peter Shaw) *A New Method of Chemistry* (London 1741) 362.

12. See P.M. Heimann 'Nature is a Perpetual Worker' *Ambix 20* 1-25 (1973).

13. James Hutton *An Investigation of the Principles of Knowledge, and of the progress of reason, from sense to science & philosophy* (Edinburgh 1794).

14. Alexander Monro 'History of Anatomy', Edinburgh University Library MS Gen 578 D, 61.

15. Colin Mclaurin *An Account of Sir Isaac Newton's Philosophical Discoveries* (London 1748) 3.

16. *Medical Essays & Observations* (Edinburgh 1733-44) I, preface.

17. A.L. Donovan *Philosophical Chemistry in the Scottish Enlightenment* (Edinburgh 1975) 38.

18. See for example John Thomson *An Account of the Life, Lectures, and Writings of William Cullen, M.D.* (Edinburgh 1859) I, 118.

19. William Cullen *The Works of William Cullen M.D.* (Edinburgh 1827) I, 129.

20. See G.N. Cantor 'Henry Brougham & the Scottish Methodological Tradition' *Studies in the History & Philosophy of Science 2* 69-89 (1971-72).

21. *The Brunoniad* op. cit. (1), 22.

Figure 2. Joseph Black: etching by John Kay, 1787. The rockface carries the profiles of Black's medical colleagues at the University.

Joseph Black: An Outline Biography
R G W Anderson

At present there exists no satisfactory biography of Joseph Black.[1] What has been written deals mainly with the period prior to his appointment to the chair of chemistry and medicine at Edinburgh in 1766. I shall attempt to give in a narrative form a more balanced summary biography, at the same time trying to avoid those topics which are being dealt with by the other contributors.

It seems clear that Black's working life divides into two distinct phases, the period of his active philosophic investigations into chemistry and heat lasting from about 1750 to 1766, and the Edinburgh period of teaching and entrepreneurship up to his death in 1799. This is not an original suggestion by any means. Henry Brougham, who attended Black's last course of lectures, wrote of this in 1845:

"We knew there sat in our presence the man now in his old age reposing under the laurels won in his early youth . . . I have heard the greatest understandings of the age giving forth their efforts in its most eloquent tongues — have heard the commanding periods of Pitt's majestic oratory — the vehemence of Fox's burning declamation — have followed the close-compacted chain of Grant's pure reasoning . . . but I should without hesitation prefer, for mere intellectual gratification, to be once more allowed the privilege which I in those days enjoyed, of being present while the first philosopher of his age was the historian of his own discoveries, and be an eyewitness of those experiments by which he had formerly made them, once more performed with his own hands..."[2]

In writing this, Brougham was aware that the main discoveries on which Black's reputation was founded had been made thirty to forty years prior to the date of the lectures which he had attended. To use a banal expression, Black had become a legend in his own lifetime. His chemical characterisation of fixed air (carbon dioxide) in the 1750s was seen in retrospect as marking the threshold of pneumatic chemistry — his differentiation of the gas from atmospheric air was soon followed by the discovery and investigation of hydrogen, nitrogen, oxygen, chlorine and nitrogen oxides by Henry Cavendish, Daniel Rutherford, Joseph Priestley, Antoine-Laurent Lavoisier and Wilhelm Scheele respectively. His experiments in heat were followed by the development of the separate condenser steam engine by James Watt which was in widespread use in Britain by the end of the 18th century. Yet from the date of Black's appointment in Edinburgh in 1766 to the end of his life this progenitor of the chemical

revolution made little or no further contribution to philosophical chemistry. As has already been pointed out, the unknown Brunonian says of Black:

"Disputes he shunn'd, nor car'd for noisy fame;
  And peace forever was his darling aim."[3]
The reason for this sudden change in Black's aspirations remains unresolved.

Joseph Black was born on 16 April 1728 in Bordeaux, the ninth child of fifteen of John Black, an expatriate Ulster wine merchant and Margaret Black, a Scot. A search for evidence of his baptism in the Chartrons suburb of Bordeaux has been unsuccessful[4]; it seems that the christening was arranged privately and hence does not appear in the church records. Black's father was a significant figure in the local community and he enjoyed the friendship of Montesquieu, a fact of which Joseph was later to be proud.[5]

At the age of twelve Joseph Black was sent to school in Belfast with two of his brothers to learn Latin and Greek. The school itself has not been identified. In 1744, when sixteen, Black entered Glasgow University to commence, as was usual, the arts course. Glasgow University, it is sometimes forgotten, is one of the ancient British universities, having been founded in 1451. The High Street building in which Black would have studied was built between 1631 and 1660. It was shamefully demolished in 1870 to make way for a railway goods yard which still occupies the site.

After four years at Glasgow John Black wrote to his son urging him to adopt study which would lead to a profession. He chose medicine. Now although medical degrees had been awarded since the fifteenth century and there were nominally two professors of medicine, Glasgow University functioned as a regional university whose teaching largely centred on the liberal arts. Although the professor of anatomy Robert Hamilton taught, his medical colleague John Johnstoun did not. A medical school, in the contemporary Edinburgh sense, could not be said to have existed. But the important factor in tracing Black's education is that William Cullen had been appointed lecturer in chemistry at Glasgow University a year before Black started his medical studies. Cullen himself had studied at Glasgow before journeying to London in 1729 where he probably attended public lectures in natural philosophy. Later he served as a ship's surgeon, returning to Scotland in 1732 when he practised as a physician. From 1734 he studied medicine at Edinburgh and eventually, six years later, graduated M.D. at Glasgow. There Cullen started teaching, initially extramurally, but in 1746 he gave a course within the University. A year later a lectureship in chemistry was established, using money saved by postponing the installation of the professor of oriental languages. Cullen was appointed and was undoubtedly the major influence on Black's student career at Glasgow. Cullen was the first teacher of chemistry in Scotland to extend the scope of the subject beyond its application to pharmaceutical preparation. He was interested in tackling problems in such areas as agriculture, bleaching and brewing and this, to some extent, was reflected in his course. Black recorded in an auto-biographical fragment which survives:

"Dr Cullen about this time began also to give lectures in chemistry which had never been taught in the University of Glasgow and finding that I might be usefull to him in that Undertaking he employed me as his assistant in the laboratory."[6]

Thus at an early stage in his career as a medical student Black was plunged into laboratory work with a teacher who shunned the narrow approach of pharmaceutical chemistry and who was closely involved in the improvement of chemical processes in emerging Scottish industries.

In 1752 Black left Glasgow for the University of Edinburgh, almost certainly because of the high esteem in which the medical school was held. It had been set up in 1726 partly for economic reasons: to make it unnecessary for Scots to travel abroad for their medical education and to attract foreign students to Edinburgh. In this latter aim it was successful and it rapidly gained a reputation throughout Europe. The College had been established in 1583 on a site close to the southern boundary of the city. Close by, in Black's time, were the Royal Infirmary, an ambitious building of 1741, and Surgeons' Hall of 1697, home of the Incorporation of Surgeon-Apothecaries. Three houses in the north-east corner of the College precinct (adjacent to the physic garden) were also of significance: Black may have attended Andrew Plummer's chemical lectures which were given in the most easterly of them. Plummer, who was one of the four founder-professors of the medical faculty, had been teaching for twenty seven years and was by this stage lecturing with little enthusiasm. He had already unsuccessfully tried to transfer his duties and his reputation was low. His course consisted nearly entirely of pharmaceutical preparation demonstrations. After the inspiration of Cullen, Plummer must have appeared very dreary to Black.

Black's main task at Edinburgh was to produce a thesis to gain his M.D. degree. However he delayed settling down to his researches for a number of reasons, amongst which he listed "the shop, infirmary and private patients."[7] His thesis, when he eventually made time for this work, was based on an interest in the chemical properties of lime water which was then used medically to dissolve urinary calculi.

There is evidence that Black's earliest research, perhaps started at Glasgow, may have been on chalk and lime. How-ever as the medical efficacy of quicklime prepared from different sources was the subject of a vigorous dispute between the Edinburgh professors Charles Alston and Robert Whytt, Black judiciously chose to base his research on magnesia alba, a basic magnesium carbonate. He found that on heating this substance a gas was expelled which he suspected had originated in the pearl ashes, or potassium carbonate, used in its preparation. By means of a cyclic scheme of quantitative experiments, the first of its kind, Black showed that the original weight of magnesia alba could be recovered by dissolving the product of heating, magnesia usta or magnesium oxide, in sulphuric acid and reconstituting the magnesia alba as a precipitate by the additon of fixed alkali, suggesting that the origin of the gas, fixed air, was indeed the alkali. The methodology devised by Black was quite original, but because the work was being submitted for a medical degree he felt obliged to include in his thesis a discussion on the effect of magnesia alba on the acidity of the stomach and as a purgative, but he later discarded this relatively trivial part. The work was finally published in June 1754 as *De Humore Acido a Cibis Orto et Magnesia Alba* (on the acid humours arising from food, and magnesia alba). It was dedicated to Cullen and Black immediately sent him a dozen copies, enquiring whether he thought it worth publishing in a journal.[8]

From the date of submission until mid-1755 Black continued his quantitative research on the reactions of alkalis and in June he read to the Philosophical Society of Edinburgh a revised account of the experiments which formed his thesis, together with the results of further investigations into chalk and lime. This paper was published in the following year and was to be his most significant publication. He had demonstrated the relationship between quicklime, fixed air and chalk and he indicated that fixed air was chemically distinct from atmospheric air.

Throughout this and later work Black used simple apparatus. Indeed he seems to have taken a pride in using the simplest of available utensils. Though he did not describe the all-important balance in his paper, there survives in the Playfair Collection at the Royal Scottish Museum a balance, long known as Black's balance, presented in 1858.[9] Though its certain provenance cannot be traced earlier than this there was by this date a tradition that it had belonged to Black and it certainly came to the Museum from the university chemistry laboratory. Its degree of delicacy is commensurate with Black's recorded weighings.

In 1755 Plummer suffered a stroke and was prevented from teaching. Black was immediately available to substitute for him and he started the annual chemistry course in Plummer's laboratory in October. This was only a temporary expedient, however, and a month later Cullen was appointed conjoint professor with the still incapacitated Plummer. Plummer himself was not consulted in this matter. Members of the Senatus Academicus raised a furore over what they considered to be high-handed and indelicate tactics of the Town Council who, as patrons of the

University, had made Cullen's appointment. Thus when Cullen arrived from Glasgow to start teaching in January 1756 he was not granted the use of Plummer's own extensive laboratory facilities but had to lecture in the inferior, adjacent laboratory of a manufacturing druggist, James Scott. On Cullen's move to Edinburgh, Black was appointed professor of anatomy and botany, and lecturer in chemistry, in Glasgow. In the following year he was made professor of medicine. Though immediately burdened with administrative duties, he continued to experiment on alkalis and started his first chemistry course in 1757 which he continued to teach for ten sessions. Black was very dissatisfied with his laboratory, however, complaining that it was damp, that the floor had never been laid nor the walls plastered and other such matters.[10] An acrimonious dispute followed with the Principal who objected to expenditure on adapting the proposed building which he called "spacious and ornamental".[11] Black got his way in the end, though this was the first of many problems he had with his laboratory accommodation throughout his career.

Early in his period of teaching at Glasgow Black came into contact with James Watt (the "mathematical instrument maker to the University") who supplied him with laboratory apparatus and other materials.[12] Eventually they went into partnership. Black's major researches were into the theory of heat, developing concepts of latent and specific heat. These endeavours are being dealt with elsewhere; suffice to say that Black started teaching his ideas at Glasgow, and though strongly urged to publish these on a number of occasions, he showed a disinclination to do so.[13] However a pirated edition of his lectures on heat (together with the sections on chemical apparatus) was published in 1770.[14] The reason for Black's unwillingness to publish remains uncertain.

Meanwhile in Edinburgh the health was declining of John Rutherford, one of the other founder-professors of the Edinburgh medical school. It was expected that Cullen would succeed him. However Rutherford was prejudiced against Cullen (for complex reasons, including the latter's non-Boerhaavian approach) and favoured John Gregory who held the chair of physic at Aberdeen. In the event, Rutherford's party prevailed and Gregory was appointed. By chance Whytt died a month later. Though Cullen initially decided not to advance himself as a candidate for the chair of the institutes of medicine because of what he considered to be unfair treatment by the Town Council he eventually consented, on being urged by his friends, as it vacated the chemistry chair and afforded the opportunity to bring Black back to Edinburgh.

There was great enthusiasm for Black's appointment in 1766 and we are told that most Glasgow students followed their mentor to Edinburgh. Black probably started teaching in Scott's house or possibly in an adjacent more recently built laboratory. Thus the medical school, until the last quarter of the 18th century, was concentrated on the following buildings: Plummer's laboratory, used until

1755, Scott's house (and later the laboratory) used first by Cullen and then by Black until 1777, Monro's anatomy theatre in the College proper, dating from 1764[15] and the Royal Infirmary designed by William Adam which was finished in 1741. Once again, Black was not satisfied with his laboratory facilities.[16] Their inadequacy may have been caused by his very success as a lecturer and the subsequent crowds of students who flocked to his chemistry course.

In 1777 Black was provided with a laboratory and class room in a building which also housed the natural history and mineralogy specimens but this too proved unsuitable when John Walker was elected professor of natural history two years later and arrived with his own large collections. Black at last had a purpose built laboratory made available to him in 1781; in April of that year he reported that he was "examining and flitting a vast rubbish of things"[17] prior to his move.

At this time the College consisted of a miscellaneous collection of inadequate buildings, some in a ruinous condition. The earliest, Hamilton House, dated from 1554, the latest being Monro's anatomy theatre and Black's laboratory. In 1785 an attempt to revive the plan to rebuild the college was made. Ambitious plans drawn up by Robert Adam were chosen and in November 1789 the foundation stone at the east front was laid with great ceremony. The scheme involved a chapel and a number of houses for professors but not, to Black's annoyance, one for the professor of chemistry. Black took matters into his own hands, carrying out a clandestine dialogue with Adam.[18] He suggested that the chapel be omitted as it was, as he put it "in imitation of the english and foreign colleges"[19] and he recommended that a house be built for him in its place. Adam adapted his original plan to allow Black his house and an engraving of 1791 shows it to be adjacent to the chemistry laboratory and classroom on the north east corner of the larger of two quadrangles.[20] In 1790, Black's 1781 laboratory had to be demolished to make way for the new one but in the event, with Adam's death in 1792 and money running out, the scheme ground to a halt with only two sections of the grandiose plan complete. Just where Black taught for the remainder of his career is not clear but the facilities cannot have been to a very high standard.[21]

Black's scientific work underwent a major change in character on his return to Edinburgh. Instead of pursuing the more fundamental problems of philosophical chemistry he took an increasing interest in the rapidly developing chemical-based industries in Scotland and he was financially involved in a number of them. He also became deeply involved in teaching. In a letter to Watt of March 1772 he wrote "I have no Chemical News, my attempts in chemistry at present are chiefly directed to the exhibition of Processes and experiments for my Lectures, which require more time and trouble than one would imagine"[22]; and in February 1786 he explained to Lord Dundonald that he could not visit his works at Culross because he was in the middle of teaching a course of chemistry – he wrote "there are duties we dare not neglect – the students would be dissatisfied and would have a right to complain."[23] His degree of devotion to his medical students appears not entirely unambiguous if the remark made by Sylas Neville on 4 May 1775 in his diary is considered:

"This day Dr Black finished his lectures rather unexpectedly, as he had promised to give more pharmacy than usual; but instead of that he gave less, alledging as an excuse that his health required some relaxation. But I cannot help thinking that abridging his proper course after beginning another at an early hour to a set of lawyers &c has not the best appearance."[24]

Black was widely consulted on many topics by those involved in commerce and industry and he undertook experimental work on the investigation of problems which arose from these matters. He gained a high reputation for his advice: a work by John Dalrymple of 1784 on the interrelationship between tar and the iron trades was based on Black's opinions and Dalrymple described Black in his book as "The best judge, perhaps in Europe, of such inventions."[25] Though it cannot be said that he made a sustained or, indeed, important contribution in any particular field, the three industrial areas which most involved Black were bleaching, furnace construction and glass manufacture. He devised an experimental arrangement for use in bleaching[26] and contributed a significant appendix to Francis Home's *Essays on Bleaching*.[27] Associated with this was his analysis of various kelps from different Scottish sources to determine their alkali content.[28] For this work he was awarded a premium by the Board of Trustees for Manufactures, Fisheries and Improvements in Scotland.

Black was greatly interested in the design of furnaces for both laboratory and industrial operations. He designed a portable, coal-burning laboratory furnace in the 1750s and later arranged for its manufacture commercially. The temperature (it was claimed) could be finely controlled by adjusting the draught by the withdrawal and replacement of brass plugs. Characteristically he did not publish details though this was done for him in 1782 by one of his German pupils.[29] No original example appears to have survived though a reconstruction has been made by The Science Museum, London.[30] The design of Black's furnace was remarkably long-lived – it was still being advertised in a manufacturer's catalogue as late as 1912.[31]

Black advised on furnaces used in glass-making, discussing their construction with his brother who was Secretary to the British Plate Glass Manufactory.[32] He was also closely involved with the glassworks of Archibald Geddes at Leith (the seaport of Edinburgh) in which he made a substantial investment. Possibly Black's involvement arose from self-interest – there is evidence that he had some difficulty in acquiring laboratory glassware. Geddes, who originally had been his pupil, became one of Black's closest friends and was appointed an executor of Black's estate. It is probable that Black was instrumental in encouraging the diversification of the products of the Edinburgh and Leith Glasshouse which at first simply produced green bottles. An advertisement for a Nooth's apparatus used for the production of artificial aerated mineral waters dates from 1787[33] and an advertisement in the 1792 *Edinburgh Pharmacopoeia* (Black served on the Royal College of Physicians of Edinburgh revision committee which produced this) mentions that standard glass measures for dispensing drugs could be purchased from the glass-house. It is fairly certain that items of glassware in the Playfair Collection, long known collectively as 'Black's Glass', were made in Geddes' works.[34] The vessels are somewhat crude and look to be the products of a bottle works rather than of a specialist producer of scientific glassware.

Finally I must turn to the area in which Black achieved his most widespread fame in his lifetime – his teaching and specifically his use of the lecture demonstration. This latter technique was, of course, nothing new, being the standard method of teaching science in the 18th century. It is known that Black's predecessors, Crawford, Plummer and Cullen used demonstrations, though it is likely that they were far more limited in scope than those which Black developed. Black transformed the chemistry demonstration into a *tour de force.* The chemistry professor at Edinburgh was unsalaried, deriving his income from students' fees. Black was by no means averse to using his skill as a showman to draw in the crowds for financial motives. Brougham, who attended the 1796 course, remarked:

"In one department of his lecture he exceeded any I have ever known, the neatness and unvarying success with which all the manipulations of his experiments were performed. His correct eye and steady hand contributed to the one; his admirable precautions, foreseeing and providing for every emergency, secured the other. I have seen him pour boiling water or boiling acid from a vessel that had no spout into a tube, holding it at such a distance as made the stream's diameter small, and so vertical that not a drop was spilt . . . The long table on which the different processes had been carried on was as clean at the end of the lecture as it had been before the apparatus was planted on it. Not a drop of liquid, not a grain of dust remained."[35]

The fame of Black's courses extended over Europe, North America and the West Indies, attracting an international audience. Each year, from mid-November to mid-May he lectured five times a week, giving about 128 lectures in all. Those attending were by no means only those intending to graduate in medicine or, indeed, those pursuing a degree at the University. The course was carefully prepared and popular. Black's lectures were divided into four main sections: the general effects of heat, that is expansion, fluidity and inflammation, the general effects of mixture, chemical apparatus and what was called the chemical history of bodies. This last and largest section was subdivided into salts, earths, inflammable substances, metals and waters. The content of the lectures was kept up to date by Black's wide circle of correspondents and his contacts with foreign students among his class, some of whom were established chemists already. It is clear that he read the journals and newly published books, introducing novel theories and discoveries into his course, though it is well known that his adoption of Lavoisier's 'new system of chemistry' was not immediate and that there were aspects which always worried him. Black devoted several lectures to the developing concepts of chemical affinity and elective attraction to explain reactions in an equation-like form and he devised diagrams to represent attraction between substances.[36]

Black's health, never strong, started to wane in the mid 1790s and he chose as his assistant and successor his former student, Thomas Charles Hope. Three years after having given his last course of lectures he died, on 6 December 1799. His imposing funeral was attended by civic and university representatives, and students. He is buried in Greyfriars churchyard, close to where he taught for so many years. His colleague, John Robison, produced an edition of his lectures in 1803 later published in America and Germany. Another colleague, Adam Ferguson, wrote in an obituary notice:

"So ended a life which had passed in the most correct application of reason and good sense to all the objects of pursuit which Providence had prescribed in his lot".[37]

To a younger generation, however, Black seemed to belong to a past era. Henry Cockburn who, as a child, had seen Black, recorded:

"He was a striking and beautiful person; tall, very thin and cadaverously pale; his hair carefully powdered, though there was little of it except what was collected into a long, thin queue; his eyes dark, like deep pools of pure water ...

The general frame and air were feeble and slender. The wildest boy respected Black. No lad could be irreverent towards a man so pale, so gentle, so illustrious. So Black glided through our rather mischievous sportiveness, unharmed."[38]

NOTES AND REFERENCES

1. Much biographical material and a discussion of Black's early career have been brought together in A.L. Donovan *Philosophical Chemistry in the Scottish Enlightenment* (Edinburgh 1975). See also Henry Guerlac's entry for Black in Scribner's *Dictionary of Scientific Biography* volume 2 (New York 1970) 173-183.

2. Henry, Lord Brougham *Lives of Philosophers of the Time of George III* (London and Glasgow 1855) 21.

3. C.J. Lawrence 'Joseph Black: the natural philosophical background', this volume, pp. 1-5.

4. Henry Guerlac 'Joseph Black and Fixed Air' *Isis 48* 124 (1957).

5. Edinburgh University Library MS Gen 874 volume 5 f11 (letter, Black to John Black, 2 September 1755).

6. Edinburgh University Library MS Dc.2.76[8]*.

7. John Thomson *An Account of the Life, Lectures, and Writings of William Cullen, M.D.* (Edinburgh 1859) I, 574.

8. *Ibid.* 50.

9. R.G.W. Anderson *The Playfair Collection and the Teaching of Chemistry at the University of Edinburgh 1713-1858* (Edinburgh 1978) 73.

10. Glasgow University Archives GUA 26643 f5.

11. *Ibid.*

12. P. Swinbank 'James Watt and His Shop' *Glasgow University Gazette 59* 4 (1969).

13. See, for example, letters from Martin Wall (15 November 1780), George Buxton (17 November 1788) and Jan Ingenhousz (26 May 1791), Edinburgh University Library MSS Gen 873 volume 1, f103; volume 3 f114 and f205 respectively.

14. *An Enquiry into the General Effects of Heat: with Observations on the Theories of Heat and Mixture* (London 1770).

15. For a plan showing these buildings, see Anderson *op. cit.* (9),6.

16. For a more detailed account of Black's laboratory accommodation, see Anderson *op. cit.* (9), 22.

17. Edinburgh University Library MS Gen 873 volume 1 f109.

18. MS letters in private collection, see Anderson *op. cit.* (9), 30, notes 62-64.

19. Edinburgh University Library MS Gen 873 volume 1 f175F (draft letter, Black to Adam, 30 December 1789).

20. Engraving 'Plan of the Principal Story of the New Buildings for the University of Edinburgh' signed 'Robert Adam' and 'Harding Sculpt 1791'.

21. Between 221 and 262 students were registered for the chemistry course between 1794 and 1799 so that a large classroom would have been required. In 1805 the accommodation was described as as "an indifferent room for a laboratory ... the furnace conveniences were very limited", see George P. Fisher *Life of Benjamin Silliman* (London 1866) I, 164.

22. Eric Robinson and Douglas McKie *Partners in Science* (London 1970) 28, letter 23 (Black to Watt, 22 March 1772).

23. Edinburgh University Library MS Gen 873 volume 4, f29 (letter, Black to Archibald Cochrane, Earl of Dundonald, 17 February 1786).

24. Basil Cozens-Hardy (ed.) *Diary of Sylas Neville 1767-1788* (London, New York, Toronto 1950) 216.

25. John Dalrymple *Addresses and Proposals on the Subject of Coal, Tar and Iron Branches of Trade* (Edinburgh 1784).

26. Edinburgh University Library MS Gen 873 volume 1, f8.

27. Francis Home *Experiments in Bleaching* Second edition (Dublin 1771).

28. Andrew Fyfe 'On the Comparative Value of Kelp and Barilla' *Trans. Highland Soc. 5* 29 (1820).

29. August Christian Reuss *Beschreibung eines Neuen Chemischen Ofens* (Leipzig 1782).

30. Department of Chemistry, The Science Museum, London SW7 2DD; the furnace bears the Museum inventory number 1977-529.

31. John J. Griffin & Sons Ltd. *Griffin's Chemical Handicraft* Eighth edition (London [1912]) 204.

32. Douglas McKie and David Kennedy 'On some letters of Joseph Black and Others' *Annals of Science 16* 129 (1962).

33. *Directions for preparing Aerated Medicinal Waters* (Edinburgh 1787).

34. For evidence, see Anderson *op. cit.* (9), 142-147.

35. Henry, Lord Brougham *op. cit.* (2), 20.

36. M.P. Crosland 'The Use of Diagrams as Chemical 'Equations' in the Lecture Notes of William Cullen and Joseph Black' *Annals of Science 15* 75 (1959).

37. Adam Ferguson 'Minutes in the Life and Character of Joseph Black, M.D.' *Transactions of the Royal Society of Edinburgh 5* (Part III) 101 (1805).

38. Henry Cockburn *Memorials of his Time* (Edinburgh 1856) 42.

Joseph Black's Work on Heat
Henry Guerlac

Joseph Black is perhaps best known for those chemical investigations on magnesia alba, quicklime, and other alkaline substances which led to his discovery of carbon dioxide gas with its distinctive properties, a gas to which he applied Stephen Hales's portmanteau term 'fixed air'. These were historic accomplishments. But I have chosen to discuss Black's other notable achievement, his work on heat, because I believe that it was quite as revolutionary as his contribution to pneumatic chemistry, and because I have never been happy with what has been written on the subject, by myself among others.[1]

We may well ask, not only how Black came to study heat and what were the stages by which he came to the twin doctrines of latent heat and specific heats, a story I shall try to reconstruct from the sparse data at our disposal, but also what relation he perceived between his work on heat and his chemistry. Let us begin with this last question. Black's *Lectures on the Elements of Chemistry*, both in the manuscript copies and in Robison's printed version published after Black's death[2], introduce the student to elementary chemistry in a manner that for his day was probably unique. Apart from some general remarks, and the kind of brief historical survey of the subject which text-book writers and lecturers felt compelled to start with, Black opened his course with a number of quite detailed and specific lectures devoted to what was then known, or believed, about heat. This, so far as I know, was quite unprecedented, and is reflected in Black's definition of chemistry. Continental chemists, in France for example, might (as Lemery did in the late seventeenth century) define chemistry as "an art which teaches the separation of different substances found in complex bodies", or later (with Macquer in 1766) as "a science whose object is to learn the nature and properties of all bodies by their analysis and their synthesis". After such a definition it was customary to proceed directly to the description of the different classes of bodies, and then take up one by one the known members of each class, telling where they are found in nature or how they are prepared, what are their characteristic properties and — very often — what were their medical or other practical uses.

Even Black's own teacher of chemistry, William Cullen, who had a keen interest in the mysteries of heat, began with

a definition (borrowed from the Dutch physician and scientist, Hermann Boerhaave, in his *Method of Studying Physik*) that can hardly be said to place the subject of heat in the foreground as Black was to do. "Chemistry", Cullen told his students, "is a science that shows the particular, as Mechanics do the general, properties of bodies".[3] And he goes on to explain that while physics or natural philosophy deals with the properties common to all bodies, chemistry treats the properties of certain bodies, or classes of bodies, properties that are peculiar to those bodies or classes. Surely this would seem to relegate the study of heat to physics, rather than to chemistry.[4]

At the outset, Black makes his position quite clear by *his* definition of chemistry, one that would surprise many of his contemporaries. "Chemistry", we read in his early Edinburgh lectures, "may be defined to be the Study of the effects of Heat & Mixture on Bodies and Mixtures of Bodies, with a view to the improvement of our knowledge".[5] Black forthwith follows this definition with as many as fourteen lectures in which he discusses "the force with which heat causes bodies to expand", and gives his version of the invention and improvement of thermometers. He asks the rhetorical question why heat, and not cold, is treated as a "positive quality", and wonders whether a body can ever be totally deprived of heat; in this connection he mentions the extreme cold recorded by Russian scientists in Siberia, and by Maupertuis and his team on their expedition to Lapland. He reminded his students that thermometers do not measure the absolute "proportions", that is the absolute intensities, of heat, "for we know not the smallest possible degree of heat", what we now call absolute zero. The coldest bodies, he remarked, must retain some heat. And he gives an account, brief but accurate, of one of the more dramatic physical discoveries of the eighteenth century, the solidification of mercury carried out by the scientists of the St. Petersburg Academy, where the mercury in the thermometer congealed, so that as Black put it, it was "like a wire flexible".[6]

In these first lectures, needless to say, Dr. Black set forth the rival theories then prevailing as to the nature of heat; the first (advanced by Francis Bacon and by Isaac Newton and most of their disciples) that heat is the motion of the constituent particles of a body; the second, which Black said most chemists favoured, was that the sensation of heat was caused by the presence or motion of a fine, particulate fluid, what Lavoisier was to call *la matière du feu* and which after 1787 came to be called *caloric*. Yet Black, although

*In this paper I have reworked material on Black as Professor at Glasgow written in the late 1950s and destined for a contemplated biography of Black. This early draft was consulted in writing my sketch for the *Dictionary of Scientific Biography* (1970).

attracted by the fluid theory, warned his students that the question "what heat is" cannot be answered. Heat can only be understood by carefully attending to the properties it manifests. This study, he insisted, "is one of the most noble & engaging in Nature", for it is heat "which animates the whole System of Nature".

One fact must be stressed before we enter on the details of our investigation: Black's major discoveries, of latent and specific heats, were not dependent upon, or deduced from, a firmly held hypothesis about the nature of heat. Nor did they originate in his own experimentation. Black never tired of repeating his distrust of *a priori* hypothetical reasoning or flights of imaginative theory. John Robison, though not to be followed uncritically, may be believed when he records that in his first conversation with Dr. Black, his teacher "gently and gracefully" checked his disposition to form theories and warned him to reject "even without examination, every hypothetical explanation, as a mere waste of time and ingenuity".[7]

Black's Move to Glasgow

In the early autumn of 1756, at the age of 28, Joseph Black journeyed to Glasgow to begin his teaching career as professor of anatomy and lecturer in chemistry, posts just relinquished by his old teacher, William Cullen, who was called to Edinburgh. Two years earlier Black had received his Edinburgh M.D. for his now historic dissertation. In 1755 he described before the Philosophical Society of Edinburgh the chemical experiments that formed the second part of his dissertation but were now considerably expanded; these results appeared the year of his move to Glasgow, in the Society's *Essays and Observations.*[8]

At the outset, Black was not happy in his professorial task; he felt competent in teaching neither anatomy nor the smattering of botany that was part of his responsibility. With surprising ease he was able to arrange with the professor of medicine to exchange chairs. The 'translation' as this odd custom was called, was formally approved by April 1757.[9] In this new post his chief duty was to deliver lectures on the institutes of medicine, which is to say human physiology. While these lectures seem to have given satisfaction, Black made no effort to be original, but "contented himself with giving a clear and systematic account of as much of physiology as he thought founded on good principles".[10] He seems never to have valued these lectures very highly, for Robison writes that he could find no traces of them among Black's papers.

Chemistry was quite another matter. He taught the subject for the first time in his second year at Glasgow, i.e. in 1757-58, or so at least he recalled long afterward.[11] At this time, and perhaps in connection with re-equipping William Cullen's laboratory, Black first met the talented young instrument-maker and mechanic, James Watt, with whom he was destined to remain to the end of his life on terms of close friendship and, during his Glasgow period, intermittent scientific association.

We have no surviving notes of Black's Glasgow lectures as we have, and in considerable number, for his years at Edinburgh. Nevertheless we can reconstruct something about these lectures and about the students who heard him. In the winter of 1761-62 one of his students was William Irvine, a favourite of great promise, who later made his mark in science. In the winter terms of 1762-63 and 1763-64, Black later recalled, his lectures had become exceedingly popular and were well attended. This was when John Robison, who returned to Glasgow at this time, after a tour of duty with the navy which took him to the siege and capture of Quebec, first heard Black lecture. He has left us a vivid description of Black's appearance and manner of lecturing which has been several times quoted.[12]

Foreign students and distinguished visitors found their way into his lecture hall. At Glasgow he had the first of those Scandinavian and other foreign students who were later suspected of carrying his unpublished discoveries concerning heat to the Continent. In the winter of 1764-5 two Russian students of some subsequent note, as the first of Adam Smith's disciples in Russia, were among his auditors.[13] In this same year he was honoured by the presence at his lectures of his brilliant new Glasgow colleague, the philosopher Thomas Reid, who had just succeeded Adam Smith as professor of moral philosophy. The letters of Reid, a luminary of the Scottish Common Sense School of philosophy, are an important and neglected source for Black's Glasgow researches, the period of his life that chiefly concerns us, for it was here that he carried out his investigations on heat.[14]

Reid followed these lectures of Black with, as he put it, "juvenile curiosity and enthusiasm", and he later was to acknowledge the great influence Black had exerted upon him.[15] The letters that Reid wrote at this time to Dr. Andrew Skene and his son Dr. David Skene, both physicians in Aberdeen, give us our first glimpse of the content and the originality of Black's lectures. If Black, as he surely did, devoted lecture time to his own discoveries on magnesia, lime and 'fixed air', it was his discovery of latent heat which seemed most striking and novel, and if Reid is a good example, captivated the imagination of his audience. Reid deemed this discovery so important that he expressed the hope (and I quote him) that Black's results "may not reach any person who be so ungenerous as to make it public before the Dr. has time to publish it himself".[16] And in giving Dr. David Skene a long account of Black's results, he stressed that he was "trusting entirely to your honour that you will be cautious not to make any use of it that may endanger the discoverer being defrauded of his property".[17]

The Discovery of Latent Heat

Despite a delicate constitution and pressing duties, these Glasgow years (1756-66) were the second great productive period of Black's research career. It was at Glasgow that he developed his doctrines of latent and specific heats and carried out alone or with his students experiments to confirm them.

Black's interest in heat, it is generally agreed, was first aroused in his Glasgow student days by the lectures and speculations of his teacher, William Cullen. While heat did not occupy the same central position that it did in Black's view of chemistry, Cullen was strongly influenced by Hermann Boerhaave, who had devoted to fire one of the longest and most famous sections of his *Elementa Chemiae* (1732), so it is understandable that expounding problems of heat was an important aspect of Cullen's lectures.[18] He was particularly interested in the temperature changes produced in what were later called endothermic and exothermic reactions.[19] John Thomson, his biographer, tells us that Cullen attempted to arrange in tabular form the variations in temperature resulting from different chemical combinations; tables, as he put it, of "heating and cooling mixtures".[20]

Cullen had begun to experiment along these lines well before his departure for Edinburgh, and had presented some of his results before the Literary Society of the College.[21] By 1754, assisted by a student, a Mr. Dobson, he made an important discovery. Dr. Cullen had put Dobson to work studying the heat or cold when various substances were added to spirit of wine (ethyl alcohol). Dobson observed that a thermometer registered a sudden drop in temperature when removed from the spirit of wine. This reminded Cullen of some results reported by Dortous de Mairan in his *Dissertation sur la glace* which suggested that cold was produced when water evaporated. Cullen discovered that repeated immersions of a thermometer in alcohol could bring the temperature down still further. He found, too, that the drop was greater and more rapid in the case of the more volatile fluids. It was clear that the cold was related to the evaporation of the film of fluid adhering to the bulb of the thermometer. To study this phenomenon *in vacuo* Cullen contrived an ingenious device to observe the changes in temperature when a thermometer was lifted out of a phial of ether immersed in a vessel of water in the receiver of an air pump. An unforeseen effect brought the experiment to an abrupt end: as the receiver was evacuated the ether boiled vigorously, producing a cold so intense that it froze the water in the surrounding vessel, although the room temperature was well above freezing.[22]

Dr. Cullen's results were communicated to Black early in 1755, and in May of that year were read to the Philosophical Society of Edinburgh. In a letter to Cullen, Black commented at the time that the paper gave him the highest pleasure and was obviously important so that new experiments should be performed to explore the phenomenon further, but that he (Black) was not then in a position to carry them out.[23] Black was clearly too preoccupied with his chemical investigations and preparing for publication his *Experiments on Magnesia Alba* to follow up this alluring prospect. But later, as Thomson suggests, Cullen's experiments on evaporative cooling played their part in turning Black's thoughts in the direction of his doctrine of latent heat. It seems, however, to have been the phenomena of freezing and fusion that really launched him into these new and turbid waters.

John Robison and Black's famous cousin, Adam Ferguson, who wrote Black's obituary, both perused early notebooks, now lost, which contained Black's first speculations about heat. They are not in precise agreement as to what these notes prove about where and when Black turned to the new field of research. Robison described the notebooks in which these important entries are made as "of an uncommon form, such as I remembered to be much used by the students of the University of Glasgow, and were sold in the shop of the University printer". On the basis of some mysterious internal evidence, Robison places these notebooks in the years 1754-5-6-7 "while", he concludes, "the Doctor was a student at Glasgow and Edinburgh, and during the first year of his Professorship in Glasgow".[24] The last statement is probably correct. But Black, of course, was not a Glasgow undergraduate in 1754 or 1755; he was at Edinburgh patching up an acceptable medical thesis, extending his chemical experiments, presenting his discoveries to the Philosophical Society of Edinburgh, and preparing for the publication of this classic paper in the Society's *Essays and Observations*. He was clearly too busy with other matters, as he suggested in the letter to Cullen, to have done much else. Moreover the argument about the student-style Glasgow notebook is hardly persuasive; professors have been known, and not infrequently, to use the same kind of notebooks as their students.

Adam Ferguson's conjectures are more credible. He gave excerpts on heat from these early notes some of which "from circumstances intermixed, appear to have been written as early as 1756", a much more plausible estimate than Robison's.[25] In one of these Ferguson found queries relating to the cold produced in the melting of snow and in the solution of salts in water. In this connection Black asked "Is it not owing to this that all bodies, in becoming fluid, have occasion for more heat than in their solid state?" And again: "Is not ice crystallized water? and does it not always feel cold, because it melts on our handling it?" This is similar to the solution of salt in water. In another notebook dating, Ferguson thought, later than 1757, Black records a curious observation made by Fahrenheit, curious at least to anybody in the mid-eighteenth century. When water is exposed to severe cold without disturbing it in any way, it can be cooled below the freezing point without congealing. But if suddenly moved or shaken it solidifies in an instant, and the thermometer rises abruptly to 32°. The act of freezing from what we call the super-cooled condition results in the emission of a quantity of heat or, as Black put it in his notes, "Is not this the heat that is unnecessary to ice?"[26] This internal heat, imperceptible to the thermometer or 'latent' must, he thought, be heat associated with the fluid state.

This observation of Fahrenheit was clearly at odds with the generally accepted view of change of state. When a solid, for example ice, is warmed to its melting point, it was generally assumed that only a small additional increment of heat sufficed to melt the entire mass. Conversely, when water is cooled to the same point, it was thought to solidify at once when a small quantity of heat is withdrawn.

Upon reflection, Black saw that the prevailing opinion was inconsistent with commonly observed, but nevertheless remarkable, facts of nature. He was particularly struck by the length of time ice or snow require to melt after the surrounding temperature had risen well above the freezing point. Were this not so, to quote John Robison, Black realized that "a fine winter day of sunshine" would at once clear the hills of snow, or a frosty night "suddenly cover the ponds with a thick cake of ice".[27] Fahrenheit's observation convinced Black that fluidity and solidification are influenced by heat in a manner very different from that commonly assumed, and that changes of state require the transfer of considerable amounts of heat: heat that is concealed in a fluid and does not act upon the thermometer.

Here is a good example of an important kind of event in the history of science. A familiar phenomenon sometimes appears in some striking or unforeseen guise: is — as it were — magnified by compression in time or space, so that it attracts the observer's attention more forcibly than its more familiar, more diffused, more commonplace manifestation. This could be, I suppose, what Francis Bacon meant by his "Striking or Shining Instances or Instances Freed or Predominant". The observation of the Dutch instrument maker was just such an enhanced or intensified expression of a common phenomenon that in its usual form attracts little attention: a 'Striking or Shining Instance'. Yet it was left to Joseph Black, decades later, to perceive its deep significance. His was the 'prepared mind'.

Without hesitation, Black included in his speculation the kindred problem of vaporization. The change of a liquid like water into a condensible vapour must have a similar explanation, as Cullen's famous experiment seemed to indicate. When a fluid is brought to its boiling temperature, a determinate amount of heat must be absorbed to effect the transformation. In this process the time factor enters in as clearly as in the case of melting snow. The everyday theory — if indeed it deserves to be called a theory, not just an uncritical assumption — required that the instant that water is brought to the boiling point, it should be transformed at once into steam. Yet any housewife could have testified that no such explosive transformation occurs; that, indeed, it takes considerably longer to boil away a quantity of water than to bring it to a boil.

In a notebook later than 1757 Black is said to have outlined a simple experiment to confirm, and to measure crudely, the latent heat of steam. This was analogous to Fahrenheit's observation on super-cooled water. Aware that under pressure water can be heated above its boiling point without turning into steam, Black proposed to place a close-corked phial of water on a stove, heat it ten or more degrees above the boiling point, and then "open it suddenly, and see how much of it converted into vapour, while the water comes down to the boiling point".[28] We shall return to this problem later on.

Black was convinced of the reality of the latent heat of *fusion* as early, perhaps, as 1756, although not on the basis of his own experiments or any quantitative evidence. According to his testimony of a later date, he presented his theory in his early Glasgow lectures in 1757-58, or perhaps the following year.[29] Black was later to record that "the importance of the surmise [that his theory of latent heat applied to the vaporization of liquids] never struck me with due force till after I had made my experiments on the melting of ice".[30] If this is the case, his early notebooks contained only casual conjectures. His experiments on melting ice, to which I shall refer in a moment, were carried out late in 1761, and soon he was telling his students about the latent heat of *vaporization* "before he had made a single experiment on the subject".[31]

Heat and its Measurement

Central to Black's grasp of the subject of heat and its measurement was his clear distinction between the *quantity* of heat in a body, and the strength or *intensity* of this quality, as he called it, in bodies: between, that is, what we distinguish as the *extensive* and *intensive* measures of heat.[32] Black presented the distinction clearly to his students in the following words:

"Heat may be considered, either in respect of its quantity, or of its quality. Thus two pounds of water equally heated, must contain double the quantity that one of them does, though the thermometer applied to them separately, or together, stands precisely at the same point, because it requires double the time to heat two pounds as it does to heat one".[33]

As commonly used for a century and a half, the thermometer measured only the *intensity* of the heat in a body, that is, its temperature. How then can heat be measured? Hermann Boerhaave in a well-known experiment (which others repeated with like results) tried to discover whether the subtile 'matter of fire' possessed detectable weight. This he did by comparing the weight of a parallelipiped of pure iron with the weight of the same mass of iron when heated to redness. His failure to detect any difference was widely quoted, although his experimental procedure was sometimes criticized. In Corollary IV to this experiment Boerhaave showed that he too understood the difference between heat and temperature:

"It were to be wished [says Boerhaave in Peter Shaw's translation], that the proportional quantity of fire contained in such a body [as the mass of iron] could be determined; but this is not so easy as at first sight may seem, by reason, though from the discovered effects of fire we may estimate its power, we cannot estimate its quantity."[34]

Black, on the other hand, found the answer. He perceived that the thermometer could not only be used to measure the intensity, that is, the temperature, but if used, as it were *dynamically*, it could determine the amount of heat transferred from one body to another. It is the *time* involved in the process of bringing bodies to a given temperature that gives a clue to the relative amount of heat lost or taken up by a body. The *quantity* of heat is related to the time of heat-flow and the temperature, as Black expressed it, "taken conjointly".

It was Newton who put Black on the right track, or rather I think it was the Scottish physician, George Martine, who in an essay first published in 1740 described in detail how Newton had used a dynamical method, and what we call his 'Law of Cooling', to estimate temperatures beyond the range of his linseed oil thermometer.[35] Newton's assumption is that the rate of cooling of a hot body placed in a stream of cold air is, at any instant, proportional to the temperature difference between the hot body and the air. While the times are in arithmetical proportion, the differences are in geometrical proportion.[36]

Experiments on the Latent Heat of Fusion

With that sense of orderly procedure that marked his work on alkaline earths and alkalis, Black saw that he must satisfy himself that a mercury thermometer is a reliable instrument by which to determine the addition and subtraction of heat in bodies. This was a necessary preliminary, much as the purification of his materials underlay the success of his earlier chemical experiments. He found with a specially built thermometer of even gauge that the expansions of mercury were "very nearly proportional to the additions of heat by which they are produced".[37]

Assured of the reliability of the mercury thermometer, Black turned his attention to measuring the latent heat of fusion of ice. A suitable method occurred to him in the summer of 1761; but, as Robison tells us, there being no ice-house in Glasgow, "he waited with impatience for the winter".[38] In December of that year Black made his decisive experiment in a large lecture hall adjoining his laboratory. His method was to compare the time required to raise to a given temperature equal quantities of ice and of water cooled to 32-33°. Both could be assumed to draw heat equally fast from the surrounding air of the room; and although the heat flowing into the ice could not be noted by the thermometer, it could be calculated from the steady rise of the thermometer in the cold water.

He took two identical, thin globular glasses (Florence flasks) and filled them both with the same quantity of water.[39] One he froze in a freezing mixture. By adding a little alcohol to the other he was able to cool it, without freezing, to the same temperature. The two flasks, supported in wire rings 18 inches apart, were warmed by the surrounding atmosphere, whose temperature during the ten hours of the experiment, remained at 47-48°. The temperature of the water rose by 7° in a half hour, while the temperature of the melting ice of course remained constant to the end. Assuming that heat entered the ice and the water at the same rate, Black calculated that the heat absorbed in melting the ice after ten hours was enough to have raised an equal quantity of water to about 140°.[40]

Black afterwards tried the experiment in another way. He suddenly applied a quantity of boiling water to an equal quantity of water frozen into ice. Instead of the resulting mixture having the mean between the freezing

and boiling temperatures, Black found the temperature to be much less. "Much of the heat", he wrote, "had disappeared in a moment; as much indeed as should have heated it to 144° more, which is nearly the same quantity as had been lost in the former experiment".[41] The average of the two experiments gives us 141°, which is equivalent to 78 cals/gram, not too far from the modern value.

Latent Heat of Vaporization

Black reported on these two experiments, outlining in some detail the theory of latent heat which they confirmed quantitatively, at a meeting of the Glasgow Literary Society. This first public account was read, according to the registers of the Society, which John Robison was able to examine, on 23 April 1762.[42]

We have seen that Black had long had "a confused idea" — these are his own words — that his theory of latent heat could be applicable to the vaporization of water,[43] and that he had proposed as early as 1757 his close-corked phial experiment, to estimate the quantity of heat which is absorbed when water is changed into steam. We know, too, that this surmise, in some form, had been presented to his students as early as 1761. But it was his successful experiments on the melting of ice, as he tells us, that led him to think that similar experiments might be successful in the case of vaporization. The experiment with the 'close-corked phial', if tried at this time, offered no prospect of yielding precise measurements, for the violence of the explosion when the phial was suddenly uncorked was accompanied by a loss of water. Black sought, rather, an experiment similar to the first one he had performed with melting ice. "The regular procedure in that case, and its similarity to what appears here in the boiling of water encouraged me to expect a similar regularity if my conjecture was well founded." The main difficulty was to procure a source of heat sufficiently regular for him to assume that the absorption of heat by boiling water could be measured by the time it was exposed to the fire. The difficulty seemed insurmountable until, as he wrote, "I was one day told by a practical distiller, that, when his furnace was in good order, he could tell to a pint, the quantity of liquor that he could get in an hour".[44]

Black sought to verify this with his own furnace by boiling off small quantities of water, and found that it was accomplished in times very nearly proportional to the quantities. He therefore set about making experiments which gave him his first values for the latent heat of steam. The first of a short series of experiments was made on October 4, 1762.[45]

The method was simplicity itself.[46] Black poured measured quantities of water into flat-bottomed tin vessels and heated them to redness on a cast-iron stove. He measured the time required to heat the water from 50°F. to the boiling point; and then the time necessary for the water to boil away. From the rate at which the temperature rose to the boiling point, he estimated that the water received 40½ degrees of heat each minute. Assuming that the heat entered at the same rate during the boiling (as when heating the water to 212 degrees) although the temperature remained constant, he computed from the average of three such experiments that the heat absorbed was equal to that which would have raised the temperature of the same amount of water, were this actually possible, to 810 degrees. This is equivalent to 450 cal/gram, compared with a modern value of 539.1 cal/gram.

These results convinced Joseph Black that his doctrine of latent heat of steam was "so completely established, that I was little solicitous of more experiments for my own conviction, or for making the doctrine clearly comprehensible to others".[47] It was not until late in 1764 — a year which James Watt made historic for European science and invention — that Black returned to the subject, in an attempt, aided by Irvine, to obtain more accurate figures for the latent heat of vaporization.

Specific Heat or Heat Capacities

Earlier in this paper I remarked that Joseph Black was the first to see the great significance for physical science of the distinction between temperature and heat, between the intensive and extensive measures of this mysterious entity. But I did not suggest what led him to this realization. I must now confront this question. The answer, I believe, is closely linked to his other great discovery about heat: the discovery that different bodies have different capacities for heat, or as we commonly say, different specific heats. Like his doctrine of latent heats, his discovery of specific heats did not arise from his own experiments, but came from his reading and his meditation. And it involved, too, like the discovery of latent heat, his encounter with a 'Striking or Shining Instance', a manifestation of the well-known fact that heat, whatever its true nature, spontaneously distributes itself among nearby bodies to bring them sooner or later to the same temperature: to reach what Black was one of the first to call an *equilibrium* of heat.[48]

But how does this equilibrium come about? Is heat or fire differently distributed among bodies when equilibrium is reached? Some persons believed that bodies absorb heat in proportion to their densities, a conjecture quite in line with the corpuscular theory of heat.

The subject greatly interested Hermann Boerhaave who recorded in his *Elementa Chemiae* some experiments that Fahrenheit carried out at his request. Some dealt with determining the temperatures that result from mixing equal quantities of hot and cold water. But one experiment that had particularly struck him gave Black an essential clue and led him to comprehend, with sudden clarity, the distinction between heat and temperature, and contributed to his second great discovery, namely that different substances have different characteristic *capacities* for heat. Fahrenheit's most unexpected result came when he mixed equal volumes of water and mercury, each at a different temperature. Surprisingly, the mercury, despite its greater density, exerted far less effect on the temperature of the mixture than did the water. This obviously challenged the theory that the heat absorbed by a body depends upon its density. Accordingly, Boerhaave concluded that this experiment showed "that fire is distributed in bodies in proportion to their bulk, or extension, and not of their density".[49]

At first the Fahrenheit experiment puzzled Black, but he soon perceived that the result showed that mercury had a smaller capacity for the fluid of heat than had a equal volume of water. He concluded that, in general, the capacity of bodies to retain heat did not vary either with their bulk or their density, but in a different fashion for which he could assign no general reason.

Once having understood Fahrenheit's experiment in this light, Black was able to see the significance of a strange experiment reported twenty years earlier in George Martine's *Essays Medical and Philosophical*. Martine had placed equal volumes of water and mercury in identical vessels before a fire, each provided with a sensitive thermometer. In repeated trials Martine found that the mercury was warmed almost twice as fast as the water, and that its rate of cooling also was much more rapid.[50] Black now saw that as less heat was required to heat mercury than water to a given temperature, the equilibrium could be reached more rapidly.

Precisely when Black came to this conclusion, and made his second great discovery concerning heat, I do not know. He was aware of the phenomenon in 1760 when he took it into account in the course of his experiments on the linear expansion of mercury thermometers. Yet he was already, it seems, exclusively absorbed with the more striking phenomena of change of state. There is no evidence that Black attempted, when the idea occurred to him, to measure the different 'capacities' or, as he often called them, the 'affinities' for heat on the part of different substances. His attention remained focussed for some years on change of state. Although Robison says that Black mentioned specific heats in his Glasgow lectures, I cannot recall any supporting evidence in the published *Lectures on Chemistry*, or in the manuscript versions of Black's Edinburgh lectures I have examined.

James Watt, Black's lifelong friend and informal associate at Glasgow, was the first to investigate specific heats experimentally: the first, to quote Robison, "who considered the subject steadily, and in a system".[51] And he was responsible for drawing Black's attention to the significance and practical importance of such research.

In any case, I can hardly bring this paper to a close without some reference, albeit brief and inadequate, to the scientific interaction between these two illustrious men.

Watt, Black and the Study of Heat

In the winter of 1763-64 Watt was asked by the professor of natural philosophy, Mr. Anderson, to repair the now-famous model of a Newcomen engine which belonged among the lecture paraphernalia. It was this event, as everyone should know, that led to Watt's invention, early in 1765, of the separate condenser which vastly improved the efficiency of steam engines.[52]

Having put the model in mechanical shape, Watt set about to make it operate satisfactorily. To this end, in the winter and spring of 1764, Watt was led to perform a series of important experiments. These he carried out independently, but he seems to have been in close touch with Black, for at least one of these early experiments — an attempt to measure the volume of steam produced by a given quantity of water — was contrived with Black's assistance.[53]

Watt was astonished at the large amount of cold water that was necessary to condense the steam in the cylinder of a Newcomen engine and produce the vacuum necessary to operate it. To obtain quantitative data he made an important experiment in the summer of 1764: he passed steam from a tea-kettle (the only tea-kettle we can confidently associate with the name of James Watt!) by means of a bent glass tube into a cylinder of cool well-water. When the water had been brought to a boil and could condense no more steam, it had increased by about 1/6 of its original quantity. In other words, water in the form of steam could heat about six times its own weight of well water to 212°F.[54]

Watt found this result incomprehensible until — upon mentioning the fact to Black — he first heard about the phenomenon of latent heat. This, of course, was many months before Watt devised his famous separate condenser, and Black may be pardoned for believing that this disclosure "contributed, in no inconsiderable degree, to the public good, by suggesting to my friend Mr Watt . . . his improvements of this powerful engine".[55] This claim was advanced even more strongly by John Robison when in 1803 he dedicated to Watt his edition of Black's *Lectures*, for he addressed Watt as "Dr. Black's most illustrious Pupil" and spoke of "your improvements on the steam engine, which you profess to owe to the instructions and information you received from Dr. Black".[56] Watt, who remained intimately attached to Black and Robison, was nevertheless later impelled to point out that he had not been a pupil of Black in the formal sense, never having attended his or any other lectures while in Glasgow. And he insisted that the invention of the separate condenser was not suggested by his learning the doctrine of latent heat.[57] He gave, nevertheless, as my readers will see before I conclude, a judicious and persuasive account of his debt to Black.

If, as I believe we must now admit, Black's disclosure of the doctrine of latent heat was not a decisive factor in the invention of the separate condenser[58], Watt's influence on the later experiments of Black and his students is undeniable. His ingenuity and questioning mind enormously stimulated the older man. It was precisely in these years 1764-5 that Black, always plagued by ill-health, began to enlist the assistance of his abler students in order to husband his own limited energy; and it is significant that the problems which they investigated were ones brought to their attention by the problems that Watt then faced. During 1764 Watt was in constant touch with Black and his collaborators. It is

doubtless from them that he learned about Cullen's demonstration that water and other liquids boil *in vacuo* at very low temperature: water, indeed, well below 100°F. With this piece of information Watt understood why the water injected into the heated cylinder of his engine did not produce an effective vacuum.[59] When we find John Robison put to work by Black in 1764 determining the temperatures at which various liquids boil *in vacuo*, the connection with Watt's investigations is manifest.[60]

An equally apparent connection with Watt's interests emerged when Black assigned his able assistant, William Irvine — who had been measuring the latent heat of fusion of spermaceti, beeswax, and various metals — the problem of determining by a new method and with greater precision the latent heat of steam. Irvine, as we saw, had been Black's chemistry student in 1761-2 and was now a candidate for a medical degree. Black described him as "a young gentleman of an inquisitive and philosophical mind, of great ingenuity, and peculiarly qualified . . . by the habits of mathematical study, and scrupulous attention to all kinds of measurement"[61] to carry out such work.

Black had come to see that a careful measurement of the latent heat of steam should be made "because this appeared the most accurate way for obtaining measures of the heat produced by fuel, and might be of great service in making trials of the most economical methods of applying fuel, a point of immense consequence in nine-tenths of all our manufactures". But ill-health, "the occupations of a double academical duty, and the attention due to my patients", stood in the way. Now that Watt's queries and experiments had opened up a new industrial application for the theory of latent heat of vaporization, Black determined to carry out with Irvine such improved measurements. The first experiment was made on October 9, 1764.[62]

The method was straightforward, and made use of a common laboratory still as a calorimeter.[63] A measured quantity of water, initially at room temperature (52°F), was distilled; as the water condensed it was collected and carefully measured. The condenser ('refrigeratory') contained a known amount of water, and the rise in its temperature was followed throughout the experiment. From a temperature of 52° at the beginning, it had risen to 123° by the time three measures of water had been collected. An easy calculation showed that the latent heat of steam — the heat required to convert three measures of water into steam — was equal to that required to raise the same quantity of water to 739°F. When allowance was made for the fact that the water which emerged from the still was 11° hotter than the water in the condenser, and that considerable heat had been lost from the surface of the apparatus, Black and Irvine estimated that the latent heat of steam could not be less than 774°. This value was far too low.

A short time after — that is to say, in October or early November 1764 — Watt repeated the experiment with a smaller still better suited to the investigation and obtained a higher value, but one that was still too low. Watt later reported improved values.[64]

Heat Capacities or Specific Heats

Special interest attaches to the measurement made at this time by Watt and by Black and Irvine. It is clear that Black was aware, as early as 1760, of the phenomenon of differing heat capacities. In determining the reliability of mercury thermometers in measurements of heat — measurements carried out by mixing quantities of hot and cold water — Black sought to avoid one serious cause of error by carefully cancelling out the effect of heating or cooling of the vessels in which the mixture was carried out. He used containers of the same materials and the same size and weight, and carried out each determination twice, once by pouring the hotter water into the colder and then *vice versa*, taking the mean of the results. McKie and Heathcote correctly, I think, saw this as an application of Black's doctrine of the different heat capacities of solids.[65]

In no aspect was Watt's questioning mind as important in Black's work as in the matter of specific heats. The earliest problem that confronted Watt in his attempts to improve the Newcomen engine — well before he hit upon the idea of the separate condenser — was to reduce the quantity of heat wasted in warming the cylinder. He tried to reduce this loss by finding metals or alloys for the cylinder which would absorb little heat. Among his earliest experiments were those intended, in effect, to determine the heat capacity of different substances, and these experiments launched Black and his students on a new experimental programme. Watt was in the habit of communicating his results to Black, who by now had completed his own pioneer researches on latent heat and who had, as we have seen, begun to assign a share in his research projects to able students and assistants like John Robison and William Irvine. It is understandable that about 1764 Black should have turned his attention to the subject that Watt had opened up, and should have carried out, at first alone, and then with the help of Irvine, numerous "experiments on the heat communicated to water by different solid bodies, and had completely established their regular and steady differences in this respect".[66] Black continued these investigations with Irvine until called to Edinburgh in 1766. After Black's departure, Irvine carried out assiduous research along the same lines, measuring the specific heats, relative to that of water, of sand, glass, iron and other substances. He attempted to measure the specific heats of gases, and tried to estimate the absolute zero of temperature. His work was never published in his lifetime.

We may conclude with the final appraisal that James Watt made of his relationship, and his debt, to Joseph Black. Late in life he wrote:

"Although Dr. Black's theory of latent heat did not suggest my improvements on the steam-engine, yet the knowledge upon various subjects which he was pleased to communicate to me, and the correct mode of reasoning, and of making experiments of which he set me the example, certainly conduced very much to

facilitate the progress of my inventions; and I still remember with respect and gratitude the notice he took of me when I very little merited it, and which continued throughout my life".[67]

To John Robison, Watt wrote on hearing the news of Dr. Black's death in 1799:

"Like you, I may say, to him I owe in great measure my being what I am; he taught me to reason and experiment in natural philosophy, and was always a true friend and adviser".[68]

NOTES AND REFERENCES

1. Ernst Mach *Die Principien der Wärmelehre* (Leipzig 1896) 153-81; Douglas McKie and Niels H. de V. Heathcote *The Discovery of Specific and Latent Heats* (London 1935) 1-53; and more recently my article on Black in the *Dictionary of Scientific Biography*, reprinted in my *Essays and Papers in the History of Modern Science* (Baltimore & London 1977), where Black's work on heat is summarized on pp.293-97. See also A.L. Donovan's *Philosophical Chemistry in the Scottish Enlightenment* (Edinburgh 1975); his discussion of heat in Chapter 9 is helpful.

2. Joseph Black *Lectures on the Elements of Chemistry*, ed. John Robison, 2 vols. (Edinburgh 1803). Robison summarized Black's contributions to the study of heat in his 'Editor's Preface', discussed it in a long and often polemical note appended to the first volume, and set forth Black's experiments and doctrine, presumably in the discoverer's own words, in the text of the *Lectures* themselves. Henceforth to be cited as *Robison*.

3. Notes of Dr. Cullen's Lectures on Chemistry made by Dr. White of Paisley, 1754, p.1. In the Clifton College, Bristol, MS 'Lectures on Chymistry by Dr Willm Cullen, Med. & Chem. Prof.', the definition is found in Lecture 2, pp.3-5, the first lecture being devoted to the definitions given by other chemists.

4. Hélène Metzger pointed out that if fire was conceived of as a material ingredient, entering into the very substance of chemical reactants, then it deserved its place in a work of chemistry: "si non, comme le déclarent les stahliens, le feu n'est qu'un instrument utile et nécessaire à la chimie, mais aucunement réactif ou ingrédient, son etúde approfondie appartiendra à la physique et les chimistes éviteront même de l'aborder." See her *Newton, Stahl, Boerhaave et la doctrine chimique* (Paris 1930) 9. This can apply only to the early followers of Stahl, for by the mid-eighteenth century in France phlogiston had come to be identified with the 'matter of fire'.

5. Lecture of November 17, 1766, Blagden MS 'Notes of Dr Black's Lectures', 1766-67, I, p.1, Wellcome Institute for the History of Medicine, London. I have also made considerable use of the Black Lectures of 1775 (Alexander Law) in the Library of the University of Edinburgh.

6. Blagden MS, I, p.18.

7. *Robison*, I, vii. Black gave Robison "Newton's Optics to read, advising me to make that book the model of all my studies".

8. *Essays and Observations, Physical and Literary, Read Before a Society in Edinburgh* 2 157-225 (1756).

9. Sir William Ramsay *Life and Letters of Joseph Black, M.D.* (London 1918) 32-33, citing a letter of Black to his father.

10. *Robison*, I, xxx.

11. Letter of Black to James Watt (15 March 1780) in James Patrick Muirhead *The Origin and Progress of the Mechanical Inventions of James Watt*, 3 vols. (London 1854) II, 119. Cf. *Robison*, I, 116: "...when I began to read my lectures in the University of Glasgow, in the year 1757".

12. *Robison*, I, lxii-lxiii.

13. Robert Scott *Adam Smith as Student and Professor* (Glasgow 1937) 158-9, and especially Professor Alekseev's 'Adam Smith and his Russian Admirers of the Eighteenth Century', *ibid.* Appendix VII, pp.424-31.

14. *The Works of Thomas Reid, D.D.*, ed. Sir William Hamilton, 2 vols. (Edinburgh 1872) I, 39-50.

15. Dugald Stewart *Account of the Life and Writings of Thomas Reid, D.D., F.R.S.E.* (Edinburgh 1802) 34 and 36; also Reid, *Works*, I, 10.

16. Reid, *Works*, I, 42.

17. Reid, *Works*, I, 44.

18. John Thomson, *An Account of the Life, Lectures, and Writings of William Cullen, M.D.*, 2 vols. (Edinburgh 1832 & 1859), I, 51 and 53.

19. Cullen's experiments and curious speculations about such reactions are described in Thomson, *Cullen*, I, 52-53, and are well summarized in Donovan's more accessible *Philosophical Chemistry*, 156-58.

20. Thomson, *Cullen*, I, 52. Cited by Donovan, p. 156.

21. On Cullen's unpublished writings on heat dating from this period, see Thomson, *Cullen*, I, 53-4 and 580-83, where one of these papers is printed *in extenso*.

22. 'Of the Cold Produced by Evaporating Fluids, *Essays and Observations, Physical and Literary, Read before a Society in Edinburgh 2* 145-56 (1756). Here Cullen's paper immediately precedes his student's 'Experiments on Magnesia Alba'.

23. Black spoke of it as "a labour in which I cannot engage myself at present, having already bestowed too much of my time on my own chemical inquiries": Thomson, *Cullen*, I, 57.

24. *Robison*, I, 525. Robison gives excerpts from six of these small notebooks. Comparison with those given by Ferguson shows that the selections made by the two men are not identical, and where they are, there are differences in transcription.

25. Adam Ferguson 'Minutes of the Life and Character of Joseph Black, M.D.' *Transactions of the Royal Society of Edinburgh 5* (Part III) 101-117 (1805). See especially p. 106. Ferguson, with scholarly caution about fixing dates except "within certain limits", nevertheless conjectures that "the notes appear to have been written while he was a student at Edinburgh, or candidate for his degree, in the year 1756". This, of course, is incorrect. Black received his degree in 1754. He left for Glasgow in the early autumn of 1756.

26. Ferguson, 'Minutes', 107. Fahrenheit's observation was reported in his paper 'Experimenta & Observationes de Congelatione aquae in vacuo' *Philosophical Transactions 33* 78-84 (1724).

27. *Robison*, I, xxxvi-xxxvii and 116-17. Robison points out (p.528) that others before Black had observed that the temperature of melting ice remains constant until all the ice has turned into water, among them Fahrenheit and Dortous de Mairan. Indeed Robison believed "that this has been noticed by every naturalist, since the time that Dr Hooke recommended melting snow for settling a point in the scale of temperatures; for the propriety of this method for ascertaining a determinate temperature rests entirely on the truth of this observation". But nobody before Black inferred from this fact "that the ice is continually *absorbing heat*, and that the heat so absorbed may be measured by the time of the liquifaction".

28. Ferguson, 'Minutes', 107. Cf. *Robison*, I, 525, where the text differs slightly, although the sense is unaffected.

29. Black to Watt, letter of 15 March 1780 in *Muirhead*, II, 119. Adair Crawford put the date of Black's discovery of latent heat as 1755 or 1756, and remarks that Black "taught it publicly in his Chemical Lectures as early as the year 1757 or 1758". See his *Experiments and Observations on Animal Heat* (2nd edition,

London 1788) 71-72, cited by McKie and Heathcote, p. 39.

30. *Robison*, I, 156.

31. *Robison*, I, xli. Robison had examined the notes taken in Black's lectures in 1761 "by a nobleman eminent for his science and learning, by which it appears that Dr Black brought his thoughts on this subject to *full maturity*, and that nothing was wanting but a set of plain experiments, to ascertain the *precise quantity* of heat which was combined with steam". The italics are Robison's.

32. This distinction was familiar in the Middle Ages and was applied, if only qualitatively, in the concept of the 'latitude of forms'. It has been suggested that, like the notion of 'degrees of heat and cold', the difference between the 'intension' or intensity of a quality, and the 'extension' or total quantity, went back at least to Galen, and was used in discussing the heat of the human body. Even the word 'temperature' has a similar medical origin. Derived from the Latin *temperatura* it originally referred, like the word 'temperament', to the balance or imbalance of the qualities of the bodily humours. See Marshall Clagett *Giovanni Marliani and late Medieval Physics* (New York 1941); and, more recently, Edith Scylla 'Medieval Concepts of the Latitude of Forms' *Archives d'histoire doctrinale et littéraire du moyen âge 48* 223-83 (1973).

33. MS lectures of 1775. See also Blagden MS, I, p.19, lecture of November 24, 1766, for a less clear statement.

34. *A New Method of Chemistry . . . translated from the original Latin of Dr Boerhaave's Elementa Chemiae . . . by Peter Shaw, M.D.* The Second Edition, 2 vols. (London 1741) I, 287. See the *Elementa Chemiae* (Leiden 1732) I, 262.

35. 'Essay on the Heating and Cooling of Bodies', dated St. Andrews 1739, in *Essays Medical and Philosophical* (London 1740) 234-35, a work reprinted as *Essays and Observations* (Edinburgh 1772) without the two clinical essays. See pp.52-53 of this more readily available edition.

36. Martine, citing Newton's *Principia*, writes that this loss of heat follows a law similar to the slowing of a body "endowed with a certain originally impressed force, and moving in a medium with a resistance always proportional to its velocity", *Essays and Observations* (1772), 53; *Essays Medical and Philosophical* (1740), 235. Although Newton's treatment about the rate of cooling struck Martine as "vastly pretty", he nevertheless wrote (p.236) that the hypothesis "is more mathematical than physical. It gives a fine and beautiful, but not a true view of nature. The heat of a body does not really decrease exactly in that proportion. For were that truly the case, the body, though continually cooling, would take an infinite time to arrive at the temperature of the surrounding medium".

37. *Robison*, I, 58-9. His method depended on the assumption that if equal amounts of hot and cold water are mixed together, the resulting temperature should be the mean of the two starting temperatures. Black's experiments, begun in 1760, had been anticipated by Brook Taylor, though he was unaware of the fact. Formulas to describe the result of mixing different quantities of water at different temperatures had been published by G.W. Krafft (1744) and G.W. Richmann (1747). There is a detailed account of their work in McKie and Heathcote, pp.55-76. For Taylor's paper see the *Philosophical Transactions 32* 291 (1723).

38. *Robison*, I, xxxvi. This account is derived from what Black told his students in the summer of 1775 (MS Lectures of 1775, I, pp.50-51) and from *Robison*, I, 119-25.

39. Here again Black is concerned, by selecting identical vessels, to cancel out the effect of the heat capacity of the glass; another illustration of his early use of his principle of specific heats.

40. "I was, at the same time", Black wrote later, "satisfied that it received the heat, because I felt the stream of cold Air, that descended from the flask, at the distance of two feet from it. Here therefore was so much heat absorbed by the melting ice, and concealed in it". MS Lectures of 1775, I, p. 51.

41. MS Lectures of 1775, I, p.52.

42. *Robison*, I, xxxviii. McKie and Heathcote reported that their efforts, and those of Professor T.S. Patterson, to locate these documents were unavailing. See *Discovery of Specific and Latent Heats*, 31, note 1.

43. *Robison*, I, 156.

44. *Robison*, I, 157.

45. *Robison*, I. 157.

46. *Robison*, I. 157-8.

47. *Robison*, I, 171.

48. Martine used the word 'equilibrium' applied to heat transfer in 1739. See *Essays and Observations* (1772), 51; (1740), 233.

49. *Shaw-Boerhaave*, I, 291. In the original (*Elementa Chemiae*, I, 270) we find: "In hoc autem Experimento quam maxime notabile habetur, quod inde mirabilis lex naturae pateat, dum Ignis per corpora ut per spatia, non juxta densitates, distribuatur."

50. *Essays and Observations* (1772), 73-78; (1740), 257-261.

51. *Robison*, I, 504.

52. The basic sources are the accounts of Black, Robison and Watt, printed *in extenso* in *Muirhead*, I, xxxv-xcii. The accounts by Black and Robison were prepared for Watt's lawyers in the patent suit of Boulton and Watt vs. Hornblower and Maberly, 1796-7. See also David Brewster's edition of *Robison's System of Mechanical Philosophy*, 4 vols. (Edinburgh 1822) II, 113-120.

53. *Muirhead*, I, lxxii and *Brewster-Robison*, II, 115.

54. *Muirhead*, I, lxxiii-lxxiv and *Brewster-Robison*, II, 116.

55. *Robison*, I, 184.

56. *Robison*, I, iii. See also Robison's 'Narrative of Mr Watt's Invention of the Improved Steam Engine', *Muirhead*, I, xli-lxvi.

57. 'Letter to Dr Brewster from Mr Watt, May, 1814', in *Brewster-Robison*, II, ix.

58. For a discussion of the evidence and a different interpretation see Donald Fleming 'Latent Heat and the Invention of the Watt Engine' *Isis 43* 3-5 (1952).

59. *Muirhead*, I, lxx; also *Fleming*, 4-5.

60. *Robison*, I, 151, where in a note Robison says "These experi-

ments were made by me in 1764 by the help of a very indifferent air-pump". Watt cites Robison as his chief source of information on this important question. See his 'Plain Story' in *Muirhead*, I, lxxxi. Watt at this time was studying the boiling point of liquids *above* atmospheric pressure.

61. *Robison*, I, 171. See also Andrew Kent, 'William Irvine, M.D.', in his *An Eighteenth Century Lectureship in Chemistry* (Glasgow 1950) 140-150; and *McKie and Heathcote*, especially pp.123-135.

62. *Robison*, I. 171.

63. A word of caution is in order about the mythical ice calorimeter of Black, referred to by such writers as Thomas Preston in his *Theory of Heat* (London 1894) 215-16, and Henry Crew, *Rise of Modern Physics*, (2nd edition, Baltimore 1935) 203. Although Black once proposed to measure the latent heat of steam, making use of his knowledge of the latent heat of fusion of ice, these plans were not followed up. Black, in fact, attributed to Lavoisier the invention of the ice calorimeter. The inventor, in fact, was Laplace. See my 'Chemistry as a Branch of Physics: Laplace's Collaboration with Lavoisier' *Historical Studies in the Physical Sciences 7* 250 (1976), note 140.

64. *Robison*, I 173.

65. *Discovery of Specific and Latent Heats*, 125. "It is clear", they wrote on another page (p.35) "that Black formulated his doctrine of specific heat in 1760 and that of latent heat shortly after this". With this sequence, I heartily disagree. It should be noted that the authors disregard the testimony of Robison and Ferguson from the early lost notebooks as well as supporting evidence placing the doctrine of latent heat as early as 1756, or 1757-8 at the latest.

66. *Robison*, I, 504.

67. 'Letter of Dr Brewster from Mr Watt, May, 1814', in *Brewster-Robison*, II, ix.

68. *Muirhead*, II, 264. I regret that when this paper was written I had not seen Arthur Donovan's article 'Towards a Social History of Technological Ideas: Joseph Black, James Watt, and the Separate Condenser' in George Bugliarello and Dean B. Doner (eds.) *The History and Philosophy of Technology* (Urbana & Chicago, Illinois, and London 1979).

Experimental Science in the University of Glasgow at the Time of Joseph Black
Peter Swinbank

"Dr. Black, a man equally philosophical in his character and in his genius, the father of modern Chemistry, though his modesty and his indolence will render his name celebrated rather by the curious in the history of that science than by the rabble of its cultivators."[1]

Reform and Regulation

The University of Glasgow, as is known everywhere except in Edinburgh, was founded in 1451, and has a continuous history since that time.[2] However, it would be surprising if the institutions established by the original Papal bull had survived unscathed by reformation, the advent of Presbyterianism, political and social upheavals and changes in educational fashion. There was indeed a major reform in 1577 when, under the charter of *Nova Erectio*, the University was extensively remodelled and the teaching put on a more secure basis. However, by the early years of the 18th century the mechanisms set up by that charter were themselves creaking, and the dissensions which seem inevitable in an academic community had reached such a pitch that

"The Colledge of Glasgou is very thin this session, and the Masters may blame themselves; their divisions and breaches have lessened the reputation of society, and multitudes nou go to Edinburgh. This moneth the Professor Simson has been very ill of his flux, which has nou continoued more than a year, and has attacked him very severly of late; yet he is some better, and teaches some, nou and then. He presses the P[rincipal] and Mr W. Anderson to teach his scholars, but yet has gote no assistance."[3]

Robert Wodrow, Minister of Eastwood (just south of Glasgow), well known for his shrewd observations and formerly University Librarian, recorded this in his diary. The breaches and dissensions which Wodrow speaks about continued through the rest of the century, so that reading the history of the University for that time seems often to be no more than a long tale of bitter litigation, but, nevertheless, in the period between 1727 and the final departure of Joseph Black for Edinburgh in 1766 the University was transformed into one of the most intellectually active institutions ever known in Scotland.

Even before Wodrow wrote his caustic comment, there were signs of change. On June 1, 1714:

"The faculty considering that it was agreed by their act that the profession of medicine be revived, that her Majesty has been pleased to appropriate forty pounds of the late King Williams gift of three hundred pounds sterling for the support of the profession of medicine in this university, and having good and satisfying information that Doctor John Johnstoun in Glasgow is a person well skilled in medicine, and very capable to teach the same, does therefore elect and present the said Dr Johnstoun to be professor of medicine in this university."[4]

And, again, on January 21, 1723:

"The Faculty appoints Mr Morthland and Mr Simson Math. Prof. or any one of them to see the instruments for Experimental Philosophy delivered by Mr Carmichael to Mr Loudon who teaches Natural Philosophy this year."[5]

Thus, studies of a medical and scientific kind were already under contemplation, or even, to some extent, in existence. And these were not purely literary studies – a point which will be returned to below. A little later, after the date of Wodrow's comment, the teaching of experimental philosophy was to be expanded:

"This day [January 6, 1726] the Professors of Philosophy gave in a list of such instruments for Experimental Philosophy as they found wanting to make the apparatus more complete, which being considered the Faculty appoints the said Professors together with Mr Robert Simson Math. Prof. or any two of them to meet the morrow at Four of the clock to concert a Letter to be sent to Mr Hauksby at London about such of the said instruments as cannot so well be provided here and to know from him the prices of the said instruments and lykewise to send a list of what further instruments he makes, to complete the apparatus, and to write also to some proper person about the prices of some Mathematical instruments which were noted as wanting and to despatch the letter by the next Mundays post at farthest."[6]

This was followed on July 8, 1726, by the report that:

"It being represented this day that Mr Francis Hauksbee at London has sent down a box with instruments towards the compleating the apparatus for experiments, being such as were most necessary at the present the price whereof according to his account amounts to Twenty eight Pounds two shillings sterling, the Faculty in the meantime do ordain a Precept to be drawn upon one or more of the Factors: and because the Professors of Philosophy have never as yet, except Mr Carmichael for his first year, made account of their stents for the keeping up and augmenting the instruments, do therfor appoint them to make their said accounts as soon as can be, that so the money in their hands of the said stents may be applied for refunding to the Ordinary Revenue what shall be paid out of if for the said use, as far as it will go."[7]

But it was not only a concerned diarist like Wodrow who was alarmed at the state of the University. Things had got so bad that a Commission was appointed and, to quote the opening words of Wodrow's account of the proceedings in Glasgow in October, 1726:
"In the end of the last moneth, and beginning of this, we had a Visitation in the Colledge of Glasgou; and a very great parade and solemnity was made. Their names, see List in Letters last moneth. My Lord Loudon and the Earl of Isla, the Lord Grame, solicitor, and the whole ministers named, though some did not stay long. They sate from Thursday till Wensday. They met but seldome, and one would think they had some other work beside the Visitation."[8]

Wodrow seems to have been distinctly unimpressed by the industriousness of the Commissioners, and it is clear from his lengthy account that the gossip of the town was to the effect that the Commissioners were much more concerned about the process of election of the Rector and with inter-professorial squabbles than with the organization and teaching of the University as it affected students. The Commission met on several occasions in Glasgow, and subsequently in Edinburgh, where it may have done most of its serious work for in September 1727 Wodrow records:
"This moneth and the last, or in June, the Commission for the Royall Visitation of the Colledge of Glasgou sat at Edinburgh. We [have] many different reports of what Regulations they have made. One good one is, that the Masters are restricted to their severall classes: Mr L[oudon?] to the Logick and Pneumatology, Mr Car[michael?] to Morall Philosophy, Mr Dick to Naturall Philosophy, and students are not oblidged to be with them all, but as they please; and may go from the Greek to Naturall Philosophy, if they please. Somewhat in the Laureations is to go to all the Regents. Another is as odd, and many think unreasonable, and contrary to the rights and common libertys of men; that none shall be Factor to the Colledge, who is a cusine-germain, or any relation nearer, to any of the Masters. Mr Dunlop is charged with bringing this in to turn out Mr Loudon's brother and Mr Carmichael's son; but he refuses this, and sayes he opposed this, and Mr Alston is said to be the mover of this in the Visitation."[9]
It is interesting to notice that concern with the rectorial election is not so prominent in this report, but the passage about the College Factor is interesting in that it shows that not merely were there quarrels amongst the masters, there were also suspicions of the possibility of irregularity in the accounts.

Wodrow makes it clear that it was rumoured in well-informed circles in Glasgow that there were to be changes enacted by the Commission to define the teaching arrangements, for he continues:
"They have ordeaned the Masters to teach, and have publick discourses, once a week, three weeks, or moneth.

That by the day they name, 1st of November, I think, or 26th of October, every Master begin to teach on his proper business; and severall other things. In short, this Visitation has laid the foundation of unhappy divisions and constant broyls in that Society, if matters be as the one side represent it. And the matter of precedency is setled; after the Professor of Divinity, all the rest are to go *secundum aetatem.*"[10]

Fortunately, Wodrow's pessimism was not completely justified! The Commission produced an Act dated 19 September 1727, which regulated the affairs of the University for the rest of the 18th and into the 19th century, providing the institutional structure within which Cullen, Smith, Black, Watt, Reid and the other stars were able to shine.[11] Most of the provisions of this Act are not relevant to our present considerations, but a number of them impinge directly on the state of science and the career of Joseph Black.

At that time a distinction was drawn between two groups of students, "gowned" and "non-gowned". The non-gowned students were not candidates for degrees in arts, but might be taking any of a range of classes. The gowned students were those who had a particular official standing within the University:
"That the time of the general matriculation of gown students shall be upon the lawfull day immediately before the election of the Rector, and that they likewise be matriculate at all other times when they require the same; . . ."[12]
This was followed by the provision
"That the students who are not gown scholars, shall, upon their being matriculated, faithfully promise to attend their respective Colleges and studies for the space of three months at least, and that they renew the like promise every year at their return to the College, before or on the day prior to the election, otherwise they are not to be admitted to vote in the election of the Rector."[13]
This matter of 'gown' and 'non-gown' students or classes is one which has produced considerable difficulty for historians in analysing the structure of courses and classes and determining the numbers of students, etc.

After making provision for the details of rectorial elections, faculty meetings, the duties of the Clerk and the care of the College property and finances, etc., the Commissioners turned to professors and teaching, beginning with the down-to-earth statements:
"PROFESSORS AND TEACHING. –
That the masters in the said University begin to teach the business of the year, in all time coming, on the 20th day of October.
That the professors of Philosophy in the said University shall, in time coming yearly, teach the several parts of their professions after-mentioned, viz. that the masters of the semi class shall teach Logics and Metaphysics, and that part of the Pneumatics de Mente Humana.
That the master of the Baccalour class shall teach the remaining parts of Pneumatics de Deo and Moral

Philosophy.

That the master of the Magistrand class shall teach and go through a course of Physics and Experimental Philosophy.

That whichsoever of the present masters shall hereafter have the teaching of the Baccalour class, he shall have six guineas yearly out of the fore-end of the laureation-money, and the remainder thereof shall be divided between him and the other two present masters who shall have the teaching of Physics and Logics."

This provision marked the end of the old system of 'regenting', under which each master or regent had taken a single group of undergraduates through the several years and classes of the curriculum. The existence of named chairs of Latin and Greek from 1706 and 1704 respectively indicates that this was the culmination of a process which had already begun, but after the establishment of those two positions the regents still had to teach the three philosophy classes in turn.

After provisions regarding succession to posts and fees, this part of the Act goes on to lay down rules concerning class meetings and confirms the choices of the three philosophy regents of the classes which they would hence-forth undertake:

"That the three Philosophy classes have each of them a different prelecting hour.

That these hours may be attended by students without gowns, provided they have passed their course, or are students of Divinity, Law, or Medicine; and that any person, not a student as said is, may attend the lessons of Experimental Philosophy without a gown.

That neither the professors of Philosophy nor Greek shall, after the 20th day of October, and during the session of the College, teach any other thing than their own proper business.

That both the classes, from which there is access to degrees, shall sit as long as the other Philosophy class. And the Commissioners having recommended to the masters of the said three Philosophy classes, to make their election which of the classes they were severally to take, and they having agreed among themselves, and Mr. Gersham Carmichael having made a choice of the Ethic class, Mr. John Loudon of the Logic class, and the teaching of the Physic class falling to Mr. Robert Dick."[15]

This passage has several points of interest. First it confirms that the gowned classes could also be attended by non-matriculated students; second, it implies that the professor of Latin was regarded as different from the others, presumably because a substantial number of students did not need his class, having enough Latin from their schools; third, that it was necessary to tie down the professors to their chosen subjects; finally, that either of the classes of moral or natural philosophy could be taken in the final year.

The Act then goes on to lay down that for the non-gowned classes there must be a minimum of five students if the class is to be given, failing that number, the professors were to lecture publicly once a week in the session.

Detailed regulations are specified for botany and anatomy, namely:

"That the professor of Botany and Anatomy teach Botany yearly from the 15th of May to the first day of July, if five scholars offer; and that Dr. Brisbane, present Professor of Botany and Anatomy in the said University, is obliged to teach Anatomy as well as Botany; and ordains him to teach Anatomy yearly, as the other professors above-mentioned are appointed to teach the business of their professions; and that he begin to teach so soon as ten scholars offer; and if no such number offer before the first day of November, that thereafter he shall prelect publicly on Anatomy once every week, as other professors are to do in the like cases, until the 15th day of May that he begins to teach Botany."[16]

These regulations show that only fairly small numbers were accepted for the non-gowned classes, and they continued in force for a considerable time.

The only other section of the Act which is directly relevant is one which concerns degrees in arts:

"DEGREES. — That students, being Scotsmen, shall not have access to the Ethic or Physic classes in the said University, unless they have been in the semi-class either in the same University or in some other of the Universities or Colleges in Scotland.

That upon their having been in a semi-class as said is, there shall be access therefrom to either of the other two Philosophy classes, viz., that where Moral Philosophy, or that where Natural and Experimental Philosophy is taught, and that there shall be access from either of the two last-mentioned classes to degrees in Arts, provided the candidates have studied the business in both these classes under the professors of these respective classes, and that they give proof of a competent knowledge in all the parts of Philosophy, and in the languages. And if they be Scotsmen who have studied no part of Philosophy out of Scotland, that they have access to degrees if they have actually studied all the parts of Philosophy in some of the Universities of Scotland."[17]

This passage was commented on by Coutts, who, writing from his great knowledge of university practice and tradition, said the last word:

"This regulation might have been clearer, but it involved that Scottish students desiring to graduate should go through the classes of Logic, Moral Philosophy, and Natural Philosophy, and should afterwards be examined in these three subjects and in languages (Latin and Greek), but attendance on classes in languages was not made compulsory. Inferences were drawn about the requirements in the case of students who were not Scotsmen, in dealing with whom the University after-wards fell into discreditable laxity, but the regulation enacted nothing positive concerning them."[18]

Students and Classes, 1746-1766

It is now necessary to investigate the structure of the University as it actually functioned in the four decades following the Act of 1727 – the period up to 1766, the year in which Black finally left Glasgow for Edinburgh.

The University records reveal a great deal about such vital matters as professorial disputes and lawsuits, the University accounts, rectorial elections and, in general, the colourful and contentious proceedings. It is fairly straight-forward to discover with reasonable certainty the names and dates of appointment of most University officials, since such appointments were usually formally made, and held until either someone else took over or death intervened, but even for the officials the details are not always easy to establish. For example, the list of Clerks of Senate published from time to time in the University Calendar had to be revised as recently as 1972, when the name of Cullen was removed and that of Black inserted. Unfortunately, though, the records are exceedingly weak in such mundane details as the students and classes which they attended. Apart from references in correspondence and minutes (which hardly contain material for a synoptic view), the principal sources of information are in the matriculation and gradua-tion registers or albums and (for the experimental classes) a limited number of inventories of apparatus and related papers. Indeed, it has long been a primary difficulty for historians interested in the University to acquire coherent data before a time well into the 19th century. One result of this has been a good deal of confusion, and even a lack of understanding, concerning the nature of the records which do exist. Another result is that even the major works of Coutts and Mackie are silent to a degree which to the uninformed seems remarkable about the day-to-day routine of the University.

After the settlement of 1727 teaching could be divided into two categories: that for the degree of M.A., and other teaching. For Scots students, as already implied, the M.A. course consisted of the classes in Latin, Greek, logic, ethics and physics, each attended for one year and normally in sequence, but not all students took the language classes, and the sequence could be varied. But, in addition, teaching was available in religious subjects, medical subjects, mathematics, law, and (from 1730) in French and even (a few years later) in Italian.

The five classes for the degree of M.A. were known as the 'gowned' classes, and students who attended them and also intended to graduate signed the matriculation album.[19] Those who attended the non-gowned classes, and probably some of those who attended the gowned classes, did not matriculate unless they wished to vote in the rectorial election. Even those students who attended gowned classes matriculated only once in their undergraduate career, and this might be in any year: those who were interested in such matters might well matriculate in an election year. It is only of the matriculated students that anything like a systematic record survives, and there is no way of knowing what proportion of the whole student body this represents – certainly not all of it. A further complication and un-certainty arises from the ruling that the philosophy classes could be attended by students without a gown, and that the class in experimental philosophy could be attended by a person who was not even a student at all in the regular way.[20] When the matriculation albums are compared with the graduation registers[21] it immediately appears that by no means all students who matriculated graduated M.A. (Joseph Black himself did not), but it does not appear that anyone who graduated M.A. had not already matriculated.

Perhaps the only conclusion can be drawn unambiguously from this is that taking the matriculation and graduation records as indicative of the state of the University is potentially misleading. Nevertheless, it is instructive and interesting to examine the records and to see what else can be learnt from them. The two years 1746 and 1766 have been chosen, for in the first of them Black matriculated in the class of ethics and in the second he moved to his Edinburgh chair.[22] The numbers of students matriculating in these years are given in Table 1.

		1746	1766
Matriculating in	Latin	17	38
" "	Greek	8	13
" "	Logic	10	13
" "	Ethics	9	18
" "	Physics	1	3
Matriculating in Gowned Classes		45	85
Matriculating for the Rectorial Election		13	47
Total		58	132

Table 1. Matriculation at Glasgow in the Years 1746 and 1766

In each of these years the largest number of matricula-tions was in the class of humanity (Latin), as might be expected, since this was the first class to be taken by students coming fresh and naive to higher education (first-year students tend to obey the rules). The sizable number matriculating in logic in 1746 may well imply (though this is far from certain) that a substantial number of students did not bother to attend the Latin and Greek classes, but passed direct to that of logic. The matriculation of a single student in physics in 1746 may represent an unpopularity of the subject: it may represent the fact that it was usually the final year subject and most students liked to matriculate before that: or it may be connected with the fact that the student and the professor shared the same surname, Dick, though they were probably not related. The overall patterns for 1746 and 1766 are generally similar, except that a smaller proportion matriculated in logic in the latter year. How far the total matriculations were affected by the rising of '45 and how far the increase over the twenty years represents a genuine increase in total University size it is hard to say.

The graduation records for the six year periods 1746-51 (in which Black's undergraduate contemporaries might be expected to figure) and 1766-71 (in which his later pupils would figure) are summarized in Table 2.

As might be expected, there are more M.A.s than all higher degrees combined, and amongst these the doctors of

1746-51	Arts	Divinity	Law	Medicine	Total Graduating	Total Matriculating
1746	10 M.A.		3 LL.D.	1 M.D.	14	(58)
1747	9 M.A.		1 LL.D.	1 M.D.	11	(71)
1748	7 M.A.	1 D.D.		4 M.D.	12	(59)
1749	13 M.A.			4 M.D.	17	(61)
1750	17 M.A.			5 M.D.	22	(57)
1751	8 M.A.	1 D.D.		3 M.D.	12	(58)
Total	64 M.A.	2 D.D.	4 LL.D.	18 M.D.	88	(364)

Table 2a. Graduation Record for the Period 1746-51

1766-71	Arts	Divinity	Law	Medicine	Total Graduating	Total Matriculating
1766	25 M.A.	1 D.D.	2 LL.D.	3 M.D.	31	(132)
1767	17 M.A.	1 D.D.	2 LL.D.	3 M.D.	23	(91)
1768	16 M.A.	3 D.D.		5 M.D.	24	(99)
1769	24 M.A.			2 M.D.	26	(98)
1770	24 M.A.			1 M.D.	25	(120)
1771	25 M.A.		1 LL.B.	3 M.D.	29	(115)
Total	131 M.A.	5 D.D.	4 LL.D. 1 LL.B.	17 M.D.	158	(655)

Table 2b. Graduation Record for the Period 1766-71 ·

medicine predominate, but the most striking feature of the table is, again, the small proportion of those who matriculated who proceeded to the M.A. degree — apparently about one-fifth in each period. Perhaps more remarkable, and certainly more significant for the present purpose, the number of M.D.s actually fell slightly between the two periods. It would be pleasant to think that the establishment of the lectureship in chemistry and the efforts of Cullen and Black led to an increase in medically-oriented study, and possibly they did, but this is not borne out by the figures. If it were possible to examine the status and education of the 18th century Scottish medical practitioners there might be a different story to tell, but the University records do not show that changes within it produced significant direct effects on the medical pro-fession at that time. It might be objected to this view that Black took numbers of medical students with him when he moved to Edinburgh, but in that case it might be expected that more M.D.s would be recorded in the period 1761-66, but the number was again 18.

There is something to be learnt from an examination of the origins, graduations and careers of a representative group of students, and it is convenient to consider those who matriculated along with Black in 1746. Of the 58 who matriculated only 15 graduated M.A. One of these, who graduated in 1747, matriculated only for the rectorial, and therefore it must be assumed was not in any of the gowned classes for 1746, though presumably the regulations were not so slackly applied that he had never attended them at all! One rectorial matriculation is recorded as having become LL.D. in 1774, but never M.A. The other M.A.s graduated at intervals between 1748 and 1754, except for one laggard who delayed until 1760. Of the matriculated students 8 became ministers of religion (including two in Ireland) and 5 of these were graduates. Seven were the sons of ministers, and of these three graduated, three became ministers but only one was both minister and a graduate. Twelve came from Glasgow, and an equal number (including Black) from various places in Ireland. This is a reflection of the fact that in Ireland the only place of higher education was Trinity College, Dublin, and in England there were only the universities of Oxford and Cambridge and all these institutions required religious tests. It is probably fair to say that the best educational opportunity for an Irishman was to come to Scotland, and, in the conditions of the mid-18th century, travel from Belfast to Glasgow by sea would be little more tire-some than travel between Belfast and Dublin by either land or sea. Furthermore, the vagueness of the 1727 Act about students who were not Scots may have made the University appear attractively flexible.

The matriculation entry for Black is worth recording in full. As already noted, the date is 1746 and it appears in the list for the ethics class:

"JOSEPHUS BLACK filius natu quartus
Joannis Black Mercatoris in Urbe
Bourdeaux in Gallia, ex urbe de
Belfast in Hibernia."

The form is, as would be expected, almost standard, though place of birth and residence are not usually both recorded.

Unfortunately, the existence of this record tells us very little about Black and the details of the course of study which he was following. He was almost certainly not in his

	Date of Foundation	1746	1766
Principal	early	Neil Campbell	William Leechman
Professors			
Latin (Humanity)	1682[a], 1706	George Ross	George Muirhead
Greek	1704	James Moore	James Moore
Logic	1727	John Loudon	James Clow
Ethics (Moral Philosophy)	1727	Thomas Craigie	Thomas Reid
Physics (Natural Philosophy)	1727	Robert Dick (the elder)	John Anderson
Mathematics	1691	Robert Simson	James Williamson
Practice of Medicine	1637[b], 1712	John Johnstoun	Alexander Stevenson (*vice. Joseph Black*)
Divinity	1640	William Leechman	Robert Trail
Oriental Languages	1709	Alexander Dunlop	Patrick Cumin
Law	1712	William Cross	John Millar
Ecclesiastical History	1716	William Anderson	William Wight
Anatomy and Botany	1718[c]	Robert Hamilton	Thomas Hamilton (*vice. Joseph Black*)
Practical Astronomy	1760	—	Alexander Wilson
Lecturers			
Chemistry	1747	—	John Robison (*vice. Joseph Black*)
Materia Medica	1766	—	William Irvine

Table 3. Teachers and Classes in the University 1746 and 1766

The dates shown are dates of institution, not necessarily of first appointment. In addition to the subjects listed, it is known that from 1730 onwards provision was made for teaching French, at first by a townsman appointed for that purpose and later by one of the professors.[23]

Notes a) The chair of humanity was founded in 1682, but lapsed in 1687, to be revived in 1706. The title of 'Professor of Humanity' is occasionally found before 1682.

b) The chair of practice of medicine was founded in 1637, lapsed in 1646, revived in 1712, endowed in 1713 and filled in 1714.

c) A lectureship in botany was founded in 1704, and the subject combined with anatomy in 1718.

first year, but more likely in his fourth year if he had studied Latin and Greek, or his second year if he had by-passed those classes. He never graduated in Glasgow at all — neither in arts nor in medicine. He did become a member of the Faculty of Physicians and Surgeons in Glasgow (M.R.F.P.S.G.) in 1757, after he had taken up the chair of anatomy — this and his Edinburgh M.D. (1754) were his academic and professional qualifications.

Before leaving the formal University records, it is convenient to enumerate the teaching staff and the dates of establishment of their chairs, and this has been done in Table 3.

This is not the place to trace Black's researches or the fine details of his University career — these are more appropriate to other papers in this symposium — but it is worth drawing attention to his appointment as Clerk. The post of Clerk to the University meeting was an important one — and it still is, under the title Clerk of Senate. Robert Simson, professor of mathematics from 1711, held the post from soon after the Act of 1727 until 1762, during which time he not only carried out the normal duties of Clerk to the University meeting but also took part in the care and organisation of the records. However, by the 1760s his powers were failing and after some difficulties[24] Joseph Black was appointed Clerk on 11th May, 1762, and held the office until 1766. During these years Black played his part in University affairs, as has been recorded by Coutts[25], and as was appreciated by the University:

"An University Meeting being duely summoned & conveened [on June 14, 1766] ...

Dr Black *in presentia*, resigned his Office as Clerk to the University Meeting, and to the other Meetings of the University, and at the same Time that the Meeting accepts of the said Resignation as a thing agreeable to the Doctor in his present Situation They appoint their unanimous Thanks to be returned to him for his faithful and gratuitous Sevices, while in that Office.

The Principal, The Dean, Dr Moor, Mr Anderson Mr Hamilton and any of the Members who may be in Town, Three of whom to be a Quorum, are appointed a Committee, to receive all the College Papers, Books &c which are in Dr Blacks possession, to lock them carefully up in the Clerk's Chamber, or other proper Repositories belonging to the College and to report to the first Meeting."[26]

The meeting then went on to consider the importance and duties of the Clerk, and arranged that in future the office should be held by the professors in annual rotation.

It is now appropriate to examine the scope and teaching of experimental and observational science in the University, as indicating an important part of the intellectual environment. It will be convenient to consider the various sciences individually, merely glancing at those in which practical work does not seem to have been a large feature but looking rather more closely at natural philosophy and chemistry.

Astronomy

As early as 1693 the University possessed an 8-foot telescope and a few other optical devices[27] but there seems no way of knowing how much these were actually used — they may have simply been used to decorate the Library. What interest there was in astronomy over the next half-century or so it is hard to determine, but on June 19th, 1752, the Faculty agreed to purchase astronomical instruments to the value of about £150.[28] Coutts records that some instruments had been acquired and that there was a proposal to build an observatory a couple of years later. But the real start in the subject came with the receipt of the bequest of instruments by Alexander Macfarlane in October 1756. The story of the way in which James Watt was paid for cleaning them has been often repeated, and so need only be mentioned, and within a few years a working observatory had been built. Initially, it was not intended to found a chair of astronomy, but simply to attach an assistant to the professor of natural philosophy. However, George II issued a Royal Warrant in January 1760, founding the chair of practical astronomy with an annual allowance of £50, and Alexander Wilson was appointed Professor and Observer. Though Wilson was a typefounder by trade he was also an M.D. and a highly competent astronomer, and by his efforts and those of his son and successor, Patrick, observational astronomy was given a sound beginning. It is noteworthy that Wilson's chair was in *practical* astronomy, for *physical* astronomy was reserved to the professor of natural philosophy. Alexander Wilson was not required to teach, but was authorized to do so. Alas, we do not seem to know with what result![29] The Wilsons held the chair between them from 1760 until 1799.

Botany

The Botanic or Physic Garden was founded in 1704, and one John Marshall, Surgeon, was appointed to be Keeper and to teach. He continued to do so until his death in 1719. Soon afterwards the combined chair of anatomy and botany was founded, and Thomas Brisbane M.D. was appointed to it.[30] Since the Royal Commission of 1727 laid down that Brisbane was "obliged to teach Anatomy as well as Botany", it appears that he did some botanical teaching, but, from what is known of his general behaviour in the University, probably not much.

After William Cullen settled in Glasgow he took an interest in botany, and at any rate by 1748 taught the subject, though Robert Hamilton had succeeded to the chair in 1742. Cullen expounded on the Linnaean system in Latin, but it is not clear what use he made of the garden or what practical work was done. However, in 1754 Hamilton and Cullen jointly pressed for improvements to the garden and the employment of a gardener, so presumably there was some actual involvement with living speciments.[31] Black succeeded Robert Hamilton in the chair in 1756, held it for only a short time and was succeeded by Thomas Hamilton, brother of Robert, who, in 1781, was succeeded by his son, William. The Botanic Garden apparently continued in a modest way and William built a conservatory in it, but it was not until 1789 that a full-time gardener was appointed. William seems to have thought that his academic colleagues did not attach sufficient significance to the garden, for in his will he tried to make provision for a new one.[32]

To sum up, it can only be said that the possibility of practical work in botany existed in the University from well before Black's time, but that it is far from established that the facilities were greatly used. In any event, it would seem fair to expect that botany would be regarded as a systematic study and as an adjunct to materia medica.

Anatomy

Parts of the story of the early teaching of anatomy bring reminders of Vesalius' condemnation of his professional brethren of the 16th century for declining the practical study of the subject. Soon after the appointment of John Johnstoun to the chair of medicine in 1717, a surgeon in practice in Glasgow was given the task of teaching anatomy within the University. This surgeon is identified by Coutts as one John Gordon, who graduated M.D. many years later.[33] After a few years it was decided to put anatomy on a more regular basis and to combine it with botany, the joint chair already mentioned was established, and the mistake was made of appointing Thomas Brisbane. But Brisbane would not dissect. In 1721 he was urged by some colleagues to teach, and the students petitioned that he should be made to do so. The point was made that in order to teach anatomy it was necessary for there to be a dissector or operator, but Brisbane's commission did not require him to operate and that therefore he was not obliged to teach! This argument is well up to the standards of the 1980s. As recorded above, the Commission of 1727 took a very different line, and laid down that he must teach. One imagines that he did not do so with any great enthusiasm. The way forward was to appoint surgeons from outside the University to do the teaching (including dissection), at first session-by-session, but later for a longer period. This odd arrangement of having a professor of anatomy and appointing outsiders to do his work seems to have functioned quite well — at least the Faculty seemed pleased with it and at least it made sure that students were instructed.

When Brisbane died in the spring of 1742 he was succeeded by Robert Hamilton who in turn was followed in 1756 by Black. They were each specifically appointed as "Professor and Teacher of Botany and Anatomy", and, since the mob on several occasions between 1744 and 1749 attacked the College on account of the work of practical anatomy, it is obvious that there must have been a certain amount of dissection.[34]

When William Hamilton held the joint chair there was certainly active anatomical study, for Coutts records that the accommodation for the subject had to be improved, and in 1776 there was trouble when it was believed that the University might be concealing the body of a suicide and keeping it for anatomical purposes.[35]

From the little evidence about both botany and anatomy the impression is gained that it was largely due to the work of Thomas and William Hamilton that the subjects were secured on a proper academic and investigative footing.

Natural Philosophy

It is not at all clear when an experimental component first appeared in the teaching of natural philosophy, but it seems to have been established by the end of the 17th century when the provision of apparatus was recognized as inadequate, but the University funds were too heavily pressed to provide more.[36] So, in about 1710 the honourable device of a public appeal was resorted to, and printed proposals were prepared (Figure 3).[37] The precise date of this document is not known — the surviving copies (two on a single sheet) may be only proofs, and it will be noted that there is a blank in the date given in paragraph IV. What is certain is that an appeal was made in terms very like those of this proposal, and that it was, to some extent, successful. Apart from their general interest and significance, several specific points emerge from these proposals. First, that there already were some instruments to hand; second, that the experiments were to be done before the students rather than by them; and third, that there was to be each year a course of experiments to which those who were not students in a regular way were to be admitted. This last feature was carried over to the report of the Royal Commission of 1727 as discussed above. It seems unlikely that the meeting proposed for October 1710 was ever held, for it was not until June 22nd 1711 that Alexander Dunlop, professor of Greek, was appointed to receive the money subscribed for purchasing instruments[38], and in August of the following year the instruments were received by the Faculty, listed, and, it would seem, handed to the care of Robert Simson, professor of mathematics and one of the users.[39] The records mention here, and in other places, the existence of an inventory, but none has come to light from so early a date. Within a few weeks — nicely before the beginning of session — regulations for the use and care of the instruments were devised and entered in the Minute Book, laying down that students in the natural philosophy class were to pay 3/- sterling for necessary expenses: but this was hardly a laboratory fee, for the regent to whom the instruments were committed was obliged to "work the Experiments himself, except when the assistance of moe hands is necessary".[40]

Over the next few years there is disappointingly little in the Archives relating to the instruments except for fairly regular records of their being transferred from one regent to another. However, interest in the collection was by no means dead, for in 1725 there was a review of the state of the instruments which led to the order to Francis Hauksbee, the younger,[41] and a few months later it was noted that the various masters had not been giving in regularly an account of the use made of the money collected from the natural philosophy students.[42] At about this time, and no doubt related to the establishment of Dick as professor of natural philosophy, a major inventory was prepared. The list begins with an air pump and its accessories, including a pressure gauge and receivers big enough to hold animals, cupping glasses, a bottle for weighing air and other instruments for pneumatics constructed in both brass and glass. It then proceeds to balances and weights, including a hydrostatic balance and various other instruments for statics and hydrostatics before going on to thermometers and barometers. Quite a few of the objects are described so briefly in this part of the list that they cannot be identified either in structure or function, but the fourth large page of the inventory gives optical apparatus, making it clear that refraction in a prism was studied, and mentioning at least three telescopes and five microscopes, as well as magnifying glasses and other optical bits and pieces. The next page goes on to globes, armillae, surveying instruments, etc. Napier's rods or "Merchistone's bones" are listed, followed shortly by "A large instrument of Wainscoat for Observing the Laws of Motion in the Collision of Bodies", and on to other apparatus for demonstrations in mechanics. From the nature of the items it is difficult to enumerate them, but there seem to be about 100 distinct groups or sets of apparatus. Taken all in all, the inventory shows clearly that there was a very substantial collection of experimental equipment in use in the University by this date. Looking through it brings memories of the contents of the back cupboards of school science laboratories of recent school days — in other words, an extensive hoard, some of the things being kept because they were merely potentially useful. The list of instruments received from Hauksbee is added to the major inventory, and at this distance in time it is very hard to see why the particular items on it were chosen, except for a group of lodestones of which the University was otherwise short.[43]

In May of 1728 proposals were considered for setting up a new experiment room, in terms which make it clear that this was to be a replacement for one which was no longer considered adequate[44] and a couple of years later proposals were to be considered for "a full course of Experiments".[45] Unfortunately, though it was agreed that a draft of these proposals should be considered next day, it was not, and no further trace of them has yet emerged. By June of 1731 the Faculty was having to agree to increase the pay of Henry Drew, who was employed to take care of the instruments and, indeed, was the first laboratory assistant in Glasgow.[46] From about this time or a few years earlier the succession of such assistants is clear. In the very early days they seem to have been paid half-a-

PROPOSALS

By the Faculty of the University of *Glasgow* for buying Instruments necessary for Experiments and Observations in Natural Philosophy.

The Faculty of the University of *Glasgow*, Considering, That Natural Philosophy as founded on Demonstration and Experiment, is justly Esteem'd one of the most Usefull, as well as Pleasant and Entertaining Parts or Humane Knowledge: But that the Tryals necessary to Illustrat that Science with the Light which it ought to receive from Experiment, Require a Collection of several Instruments and Machines, whereof this University, notwithstanding the Care by them taken to Furnish it in that kind, still wants the far greater part; For the Purchasing of which, it has at present no sufficient Fund; Do therefore find themselves obliged to apply to all Noblemen and Gentlemen who value Use-full Learning and own a concern for the Effectual Education of Youth, that they wou'd be pleas'd generously to Contribute by Subscriptions, to make up a Sum sufficient for procuring such Instruments; And if this Design meets with the Encouragement that's hop'd for, It's Propos'd,

I. That upon the Credit of the Subscriptions such Instruments and Machines as are most Necessary, shall be bought from the best Artists and brought to the College of *Glasgow*, where they are to remain and be used for the Instruction of such as shall Study Natural Philosophy and Mathematicks.

II. That the Faculty shall commit the immediat Charge and Trust of the Instruments so Purchas'd, together with those they already have, to one of their own Number, who shall be accountable for 'em to the said Faculty.

III. That it may be known, to whom this University is obliged for those Instruments; there shall be kept a Publick Register of the Benefactor's Names, and of the Sums they subscribe for.

IV. That every Subscriber shall have a Printed Inventary of all the Things bought, with their Prices; and shall be advertis'd to meet, such as please, with the Faculty upon the   day of *October*, 1710. to receive the Accompt of the Money disburs'd, and see the Instruments; and that, if it shall be found that the Money subscrib'd for, is more than what shall be instructed to be laid out in buying and bringing Home the said Instruments; The Faculty shall see the *superplus* apply'd in such manner as the Subscribers shall Determine.

V. That every Session of the College there shall be at least one Course of Experiments given to all who are inclin'd to attend it, at such times as Gentlemen of Riper Years, and others thô not Ordinary Students, as well as those that are such, may have the Benefit thereof, and Publick Intimation shall be made of the time, when it's to begin.

VI. For further Encouragement to the Subscribers, and that they may be satisfi'd of the Use of those Instruments; Any of 'em shall always be welcome to be present at the said Course of Experiments, or any part of it, as their Affairs shall allow or Inclinations lead them, and any substituted by them shall be admitted *gratis* for the first Year, and besides the Experiments in the Ordinary Course the Subscribers coming at any time to the College during the Session thereof, shall have wrought before 'em any particular Experiment they shall desire.

Figure 3. Printed proposals for purchasing scientific apparatus for the University of Glasgow, 1710. (Reproduced by permission of Glasgow University Archives)

guinea a year, but from this point onwards they had a rise to three guineas.

Over the next few decades there are various records of new instruments being bought by Dick, and in 1740 he pointed out that a sum of money had been allotted from the annual income of the University for the purchase of instruments.[47] As will be noted below, instruments for chemistry were not an established charge on University funds since chemistry was not a foundation class, but the case of natural philosophy was different. There is nothing to be gained by merely reproducing lists of instruments and repairs, but the accounts show that quite considerable sums were being spent, ranging from £38.5.6d in 1735[48] to £13.19.9½d in 1753-55.[49]

Three further documents are of considerable and pertinent interest. The elder Robert Dick was succeeded by his son, also Robert, the transfer of duties taking place in 1751-52. Each man submitted a memorial to the University about his work. The elder Dick says that between 1735 and 1751 he has spent £57.16.7d of his own money, of which £6.16.7d was for equipment and the remainder was for attendance. He goes on to say that his "profits for the great toil of the Experimental Class having been verie smal, by a smal number of students yearly applying to him, he imagines the University Meeting will think it hard he should be under so great a burden as the above sum amounts to . . .".[50] This was but one stage in negotiations between Dick and the University to get his money back! His son and successor points out that the number of students attending his class is very small and that he has no salary, and therefore feels himself justified in asking the University Meeting to grant him some yearly allowance.[51] The point here is that both Dicks had been dependent on class fees for their income, and times were hard!

The third of these documents is an account submitted by James Watt in 1759 when he was clearly employed by John Anderson, who had succeeded Dick as professor, in a fairly wholesale overhaul of the equipment. The total amount comes to five guineas, and by comparing the details with the existing inventories it might be possible to reconstruct something of the way in which the collection of instruments had developed, for it includes no less than 31 distinct items. The most fascinating one is "Brass Balls for the twirling Machine, the old ones being lost: 12/6". This account, of course, was produced soon after the time at which Watt set up his business on University premises and was given the title of Mathematical Instrument Maker to the University.[52]

Throughout the rest of the period of Joseph Black's association with the University we can see from the records that the collection of instruments for physics was kept reasonably in order and was actively used. The great uncertainty is how far they were actually handled by the students as opposed to being used merely for demonstrations.

Chemistry

There are few chemistry departments which can be said to owe their origins to a professor of oriental languages, but this was the case in Glasgow.[53] Morthland, professor of oriental languages, died on 4th September 1744, but his successor, Alexander Dunlop, delayed taking up his duties until 23rd October 1745. It was agreed that the salary so saved (£30), together with a further £22, should be devoted to setting up a chemical laboratory, and Dunlop himself formally moved on January 5th 1747 that teaching should be begun.[54] Thus, in Glasgow, a laboratory clearly intended for teaching was established from the beginning of the chemistry class.

The first teachers were John Carrick (who fell ill after delivering a few lectures) and William Cullen, and the class seemed to establish itself in University favour from the beginning. Coutts reports[55] that £136 was spent on the chemistry laboratory in 1747 and 1748, though the authority for this statement is not clear. Progress in the subject was considered in June 1749, for on 24th June the Minutes of a University Meeting record:

"Dr. Cullen having received about 2 years ago Fifty-two pounds Sterling to be laid out in building furnaces and fitting up a Laboratory and purchasing the Necessary Vessels for it, It was now reported that he had upon the twenty first current acquainted the Faculty of the Success of this attempt to begin Chemical Lessons in the University, and the University Meeting now return him thanks for the great care and pain he has been at in giving Chemical Lessons and Explaining them constantly by the most useful and necessary Chemical Processes and experiments . . ."[56]

Unfortunately, what Cullen had said on the 21st does not seem to have been recorded, but it is quite clear from this Minute that the class was regarded as a considerable success and that, as had been intended, it included experimental work.

As early as November 1749 Cullen was in process of taking over the duties of the chair of medicine, but a Minute of November 1st records that he wished to continue to teach chemistry, and on the 14th

"The Meeting being willing to allow Dr Cullen Twenty Pounds Sterling for this year to encourage him to give Chemical Lectures and Experiments, appoint Robert Simson and Mr Dunlop to acquaint him with it and to report."[57]

Again, the prominence given to experiments is seen. It is also noteworthy that Cullen was to receive a fee of £20 for a year's work. That this sum was voted each year throughout the periods of both Cullen and Black emphasizes that the lecturer in chemistry never had a permanent appointment or 'tenure' in the modern sense and this produced a problem for Black at a later stage. After the establishment of the class there was a period of development, with Black taking over from Cullen in 1756, and the laboratory continuing in existence. In the summer of 1757 Black was authorized to spend up to £40 on the laboratory.[58]

Quite soon after James Watt was given the title of Mathematical Instrument Maker to the University in 1757, he was to be found working for Black's class. For example, on October 8th 1757 he entered in his Waste Book "To the College for turning 9 doors for Furnaces for Dr. Black's class 2/-". In January of 1758 he sold to Dr. Joseph Black a condensing syringe and a "screwed head". There are various other references to repair work and sales, but unfortunately neither the records preserved in the University nor those in the Boulton and Watt Collection at Birmingham Public Library make it easy to determine in detail just what work Watt did for the class and what he did privately for Black.[59]

It does not appear that there was any kind of regular laboratory assistant in chemistry, as there certainly was in natural philosophy, but it is clear from the records that the class grew, for by 24th June 1763 the accommodation was no longer adequate and proposals were put for the construction of a new laboratory:

"The Meeting . . . are of the opinion that the Society may with prudence and safety proceed to Build a Laboratory and therefore appoint Dr Black to prepare proper plans and estimates to be laid before the Society in October next for building a Laboratory of which the whole expense shall not exceed Three Hundred and Fifty pounds Sterling.
The Meeting likewise appoint Dr Black and Dr Williamson to prepare proper Plans and estimates for fitting up the present Laboratory to serve as a Mathematical Class.
The Principal and Mr. Clow desired it to be remarked that they dissented from the above opinion and resolution.

Adam Smith,
Vice Rector."[60]

Now this represents an enormous increase in investment above the £52 originally allotted to Cullen. It might be expected, in view of the tendency of Glasgow teachers at that time to quarrel with each other, that there would be a fuss about the expense, and indeed there was.

On October 28th the Principal and Mr. Clow explained the reasons for their dissent — the two chief grounds being that the decision was unconstitutional and that there was no money. There is no need to expound their view in detail, but it was stressed that

"The College has already fitted up a Laboratory within these fifteen years at a very considerable expense which is sufficient for all necessary and useful purposes, Nor ought expenses on this article to be unnecessarily incurred, since the Profession of Chemistry has no Foundation in the College and the Society has hitherto been obliged to support the Teaching of it by an annual allowance of twenty pounds out of College Funds."[61]

The Principal and Clow go on to say that if there were funds to spare they should be spent on settling some of the University's debts. Furthermore, they claim that the present accommodation is ample. The financial objections really amount to saying that since chemistry is not an established subject, with a permanent teacher, and since there are no funds specifically for it by way of endowments, it has become a charge on the general revenue, which is already over-committed and so chemistry must be cut back. The constitutional objections are a little more subtle. Then, as now, the University had a number of "Meetings" or committees, which tended to be jealous of their own rights and privileges. It was regarded by the objectors as improper for the Meeting of 24th June, sitting under the chairmanship of the Vice Rector, to resolve to spend money in this way.

A little later, on October 31st, Black commented that the present laboratory was too small, was damp and disagreeable with no proper floor and unplastered walls. Clearly, even to make it suitable for a mathematical class was going to involve the expenditure of money.

The situation was resolved at a meeting on 2nd November, with the Rector himself in the chair, when it was agreed to build the laboratory. It must have given Black (as Clerk) great pleasure to sign that particular Minute.[62] Unfortunately there seem to be no details of the laboratory, and later references are not at all clear about it.

Black departed for Edinburgh in 1766, and there still exists the inventory prepared when he handed over the laboratory. The first part of this is worth reproducing:[63]

"Glasgow, 25 October 1766 Page first
List of Utensils delivered to the Committee appointed to receive the Laboratory from Dr. Black.

7 Distilling furnaces of white Iron each having a Grate, an Iron ring on the Top, An Iron hoop round them & a round Door of Cast Iron
A Square Door for a Brick furnace
6 Sand Pots for distilling with Retorts
An Iron Alembic to fit one of the Pots
A Copper Still with a low Copper head & a high head of white Iron
A Refrigeratory
A Copper Pan
A Papins digester of Copper with a brass lid, Iron hoop, Cross Bar & Screw
A Simple Pyrometer
A Ballance 15 Inches Beam
An Iron Mortar

Page Second
4 Supports for Receivers
A Stool for the Refrigeratory
2 drawing Boards
3 Patterns for Sand Pots
2 D⁰ for round Iron doors
A Mold for Clay doors
2 D⁰ for Muffles
A Case of Shelves for exhibiting the Salts in the order of their Composition
5 Muffles, some Clay Doors for Ashpits of furnaces & some earthen Pots for a Vent

Octr 28th 1766 Rec^d the above articles, excepting one pattern for a Sand pot.

John Robison."

This list is followed by a "List of Glasses" which records 27 different shapes and sizes of vessels, ranging from one four Quart Retort down to 128 Vials. Black had himself expended his own money on equipment, and there is a list of things which he left to be purchased:

"Page fourth

List of Laboratory Utensils which Dr Black leaves to be purchased by the College if thought proper.

	£.	s.	
An Iron Stove for the Room upstairs with a brass damping plate	2:	3:	0
The Street Crystal lamp	0:	18:	0
Iron & fixing to D^o.	0:	4:	0
A Stair Case lamp	0:	4:	0
4 tin Sconces	0:	2:	0
2 p^r Candlesticks	0:	6:	0
Snuffers	0:	0:	5
A Tea Kettle	0:	5:	3
A Chest of Small Drawers for specimens	1:	0:	0
A Cart of Bricks	0:	2:	4
Shelves with Pins for drying Vials (in the Old Cellar)	0:	10:	0
Assay furnace	0:	10:	0
	6:	6:	0
A Wind or melting furnace with brass damping plate	3:	3:	0
An experiment furnace with D^o.	3:	0:	0
	12:	9:	0

Octr 28th 1766

Rec^d the above Articles.

John Robison".

Alas, we know little or nothing of the use to which the objects in these lists were put by the undergraduates. The care and style with which they were recorded both on Black's departure and in subsequent years when the lectureship changed hands seem to imply that they were regarded as not just the lecturer's research tools.

To sum up, it is quite clear that in Glasgow the University and its various officers intended that the study of chemistry should include an experimental component, and that provision was made for a laboratory from the very beginning, and that the tradition thus established continued throughout the period which is here being considered.

Though Black was primarily professor of medicine, the University appropriately recognised his significance as a chemist in its valedictory minute:

"An University meeting being duely summoned and, conveened [on May 27, 1766] . . .
A Resignation by Dr Joseph Black of the Profession of Medicine which he holds in this University dated, the twenty fourth Instant was given in to the meeting and, accepted of,
Upon this Occasion the University are justly sensible of the loss they sustain by Dr Black's removal, his ability and, application as Professor of Medicine, his elegant & Judicious arrangements in the System of Chymistry, and the many ingenious discoveries he has made in that branch of Science, do himself and the Society great honour. His Candour, & probity of manners intitleing him at the same time to the high, Esteem and Friendship of his Colleagues.

Ja^s Clow V.R.
Joseph Black Cl. Univ."[64]

CONCLUSION

From the evidence presented here, it would seem that from well before Black's first appearance in Glasgow the University quite deliberately followed a policy of development in science, and, specifically, recognized the need for appropriate resources for observation and experiment in the various branches. The proposals for natural philosophy of about 1710 and the actual establishment of the lectureship in chemistry could both be regarded as remarkable events in an institution primarily devoted to booklearning and the pursuit of the degree of M.A. However, examination of the student records in the first part of this paper in itself demonstrates that this routine teaching was only a small part of the total activity of the University. The tendency for the professors to indulge in internal strife seems to have had remarkably little effect on the development of teaching in the scientific subjects, and even the inertia of certain professors was overcome by the vigour of Cullen, Wilson, the two Dicks, and Joseph Black.

NOTES AND REFERENCES

1. R.J. Mackintosh *Memoirs of the Life of the Right Honorable Sir James Mackintosh* (London 1835) 21.

2. The history of the University has been treated at length by J. Coutts in his *A History of the University of Glasgow* (Glasgow 1909), and in more compact form by J.D. Mackie in his *The University of Glasgow, 1451-1951* (Glasgow 1954). Many documents are printed in Cosmo Innes *Munimenta Alma Universitatis Glasguensis* (Glasgow 1854), referred to hereafter as 'Munimenta', while much diverting detail (some of it far from edifying) is to be found in the volumes resulting from the appointment of a Royal Commission on the Scottish Universities in 1826, especially the volume of *Minutes and Proceedings* issued in 1826, the *Report* of 1831, and the *Evidence . . . , Volume II, University of Glasgow*, of 1837.

3. Rev. Robert Wodrow *Analecta: or Materials for a History of Remarkable Providences; mostly relating to Scotch Ministers and Christians* (Glasgow 1842-43) III, 240; referred to hereafter as 'Analecta'. Entry dated November 1725.

4. Innes, 'Munimenta', II, 412: Glasgow University Archives (G.U.A.) 26632, June 1 1714. Before this time the University had shown some interest in medical teaching, and had even managed to confer the degree of M.D., as well as establishing a Physic Garden and a lectureship in botany. For the peculiar and even confused details, see Coutts *op. cit.* (2), 482-484.

5. Innes, 'Munimenta', II, 426: G.U.A. 22634, January 21 1723.

6. *Ibid.* 429-430: G.U.A. 22634, January 6 1726.

7. *Ibid.* 430-431: G.U.A. 22634, July 8 1726.

8. Wodrow, 'Analecta', III, 329 *et seq.* It does not seem necessary

to reproduce the list of names here: those who signed the Act of the Commission may be found in Innes, 'Munimenta', II, 581.

9. Wodrow, 'Analecta', III, 440.

10. *Ibid.* 440-441.

11. The complete text of the Act is in Innes, 'Munimenta', II, 569-581.

12. *Ibid.* 570-571.

13. *Ibid.* 571.

14. *Ibid.* 577. It is, perhaps, noteworthy that rulings under the headings Rector; Meetings; Clerk; Clerk's Chamber and Charter Chest; Estate, Chambers and Mortifications; Bursaries; Plate and Furniture; Factor and Accompts all took precedence over Professors and Teaching in the Act.

15. *Ibid.* 578.

16. *Ibid.* 579-580.

17. *Ibid.* 580.

18. Coutts *op. cit.* (2), 209.

19. For the period of interest, the albums have been published with additional biographical notes: W. Innes Addison *The Matriculation Albums of the University of Glasgow from 1728 to 1858* (Glasgow 1913).

20. This ruling may be regarded as the formal justification for the admission to classes in physics of townspeople and artisans — a development often associated with the name of John Anderson, who became professor of natural philosophy in 1757.

21. The graduation records were combined into a single alphabetical roll by W. Innes Addison in his *A Roll of Graduates of the University of Glasgow from 31st December 1727 to 31st December 1897* (Glasgow 1898).

22. The data for the tables of matriculations and graduations are derived from the works of Addison, the details of teachers from the University Calendar and Coutts *op. cit.* (2).

23. Coutts *op. cit.* (2), 230-231.

24. *Ibid.* 224-226.

25. *Ibid.* 494, and other references.

26. G.U.A. 26643, June 14 1766.

27. Coutts *op. cit.* (2), 195.

28. G.U.A. 26649, June 10 1752.

29. Coutts *op. cit.* (2), 229-30.

30. *Ibid.* 484-85.

31. *Ibid.* 489.

32. *Ibid.* 498-502.

33. *Ibid.* 484.

34. *Ibid.* 486.

35. *Ibid.* 499.

36. *Ibid.* 174.

37. G.U.A. 43105.

38. G.U.A. 26631, June 22 1711.

39. G.U.A. 26631, August 25 1712, August 28 1712.

40. G.U.A. 26631, September 15 1712.

41. G.U.A. 26634, December 30 1725.

42. G.U.A. 26634, July 8 1726.

43. G.U.A. 5191.

44. G.U.A. 26635, May 24 1728, May 28 1728.

45. G.U.A. 26635, February 16 1730.

46. G.U.A. 26647, June 26 1731.

47. G.U.A. 26648, June 25 1740.

48. G.U.A. 5194.

49. G.U.A. 5192.

50. G.U.A. 43103.

51. G.U.A. 43104.

52. G.U.A. 32201, May 23 1759. The part played by Watt in his early years in the University has been discussed in P. Swinbank 'James Watt and his Shop' *Glasgow University Gazette* No. 59, 5 (1969) and P. Swinbank 'A University Partnership' *College Courant 21*, No. 42, 3 (1969).

53. Coutts *op. cit.* (2), 239. The establishment of the lectureship is discussed by A. Kent 'The Lectureship at Glasgow' and D. Guthrie 'William Cullen, M.D., and His Times', both in A. Kent (ed.) *An Eighteenth Century Lectureship in Chemistry* (Glasgow 1950), as well as in the broader histories. Neither Kent nor Guthrie gives much prominence to the part played by Dunlop.

54. G.U.A. 26639, January 5 1747, January 28 1747, February 11 1747. The proposal as put by Dunlop explicitly recognized the importance of teaching in chemistry and the need for apparatus. It was agreed on January 5th to investigate costs, and the final decision was reached on February 11th.

55. Coutts *op. cit.* (2), 490.

56. G.U.A. 26640, June 26 1749.

57. G.U.A. 26640, November 14 1749.

58. G.U.A. 26640, June 27 1757.

59. See above, ref. (52). In addition to work connected with chemical matters, Watt and Black had a partnership with Alexander Wilson, professor of astronomy "to make some experiments on clocks".

60. G.U.A. 26643, June 24 1763. Williamson was professor of mathematics, and Clow professor of logic. From the general behaviour of the professors it seems likely that the objections were in large measure the result of personal animosities and may well have been part of a larger quarrel!

61. G.U.A. 26643, October 28 1763.

62. G.U.A. 26643, November 2 1763.

63. G.U.A. 43081.

64. G.U.A. 22643, May 27 1766.

Dr Black, a Remarkable Physician
Andrew Doig

Joseph Black's career as a physician began in 1757 when he was appointed professor of medicine in the University of Glasgow. During his nine-year tenure of this chair he was "much employed as an able and most attentive physician" and his lectures in medicine were well received.[1] He quickly gained the confidence and respect of his Glasgow medical colleagues and was twice elected President of their Faculty of Physicians and Surgeons, now the Royal College of Physicians and Surgeons of Glasgow.

In 1766, Black was appointed to the chair of chemistry and medicine at Edinburgh in succession to William Cullen who had been translated to the chair of the institutes of medicine. During the next 30 years Black continued to practise as a physician in Edinburgh though, fortunately for his health, the financial rewards of his professional activities in chemistry enabled him to restrict the size of his practice. He became President of the Royal College of Physicians of Edinburgh in 1788 and was also appointed Physician to the King in Scotland.

It is perhaps appropriate to consider some of Black's activities in academic medicine before concentrating on his abilities as a practising physician. He delivered no formal lectures in medicine during his long professorial career in Edinburgh but his lecture notes in chemistry reveal that he placed considerable emphasis on the applied aspects of his subject and especially those relating to medicine. At a time of almost universal belief in the ability of mineral springs to cure a wide range of disease, Black analysed before his class the composition of water drawn from Edinburgh's fashionable St. Bernard's Well. After discussing the results he concluded that "more service, if any, will be obtained from the exercise in walking to the Well than from the water".[2]

In the Royal College of Physicians, Black was actively involved in the preparation of the Sixth to the Eighth Revisions of the *Edinburgh Pharmacopoeia,* one of the most influential works of its kind. The 2,000 copies of the 1774 edition were sold by the following year, when permission to publish another impression was granted. The *Edinburgh Pharmacopoeia* was also printed in Germany, Holland and Italy and it played a direct role as the progenitor of several American pharmacopoeias, the first of which was prepared for the American Army in the Revolutionary War.[3]

Black's own studies on the preparation and use of drugs obtained from plants included cicuta, from hemlock.

Although it has been suggested that this substance was the Athenian state poison with which Socrates was executed[4], Black paid careful attention to the preparation and dosage of cicuta and found it of value in the treatment of whooping cough.[5] Coniine, or propyl piperidine, was found by later workers to be the major active principle of cicuta and it is of interest to note that this was the first alkaloid to be successfully synthesised in the laboratory.[6]

Joseph Black was keenly interested in the clinical effects of the medicinal gases prepared by his friend James Watt and he made a financial contribution to the pioneer work which Thomas Beddoes and Humphry Davy were carrying out near Bristol.[7] Beddoes had previously attended Black's lectures in Edinburgh and was elected President of the Royal Medical Society; Black actively supported this undergraduate society which encouraged critical debate and discussion in medicine, science and philosophy.[8]

In Edinburgh Royal Infirmary, Black served as a manager for at least four periods between 1771 and 1794, the principal duty of this appointment being to make regular inspections of the hospital and report on the findings to the Board of Managers. He himself served on the clinical staff as a physician for a period of only one month and it is often wondered why this term of office was so exceedingly brief. The explanation is recorded in the Minutes of an Extraordinary Meeting of the Managers of the Infirmary held on 25th September 1775:
"A letter was presented and read from Dr Black signifying that upon trial he found himself, on account of the state of his health, unable to undergo the labour attending the office of Physician to the Infirmary, conferred on him at the last meeting and therefore resigning that office, which resignation the Managers accept."

Black had a chronic disorder of the chest, and it is hardly surprising that his health was unable to support the extra load when one considers that it involved daily rounds of busy hospital wards in addition to his duties as professor of chemistry; he was also engaged in consultant work as a chemist and as a private physician.

Sir Walter Scott's description of his own childhood casts an interesting light on Black as a practising physician. After relating that his father and mother had twelve children and that only five survived very early youth,

Sir Walter writes:

"I was an uncommonly healthy child, but had nearly died in consequence of my first nurse being ill of a consumption, a circumstance which she chose to conceal, though to do so was murder to both herself and me. She went privately to Dr Black, the celebrated professor of chemistry, who put my father on his guard. The woman was dismissed and I was confined to a healthy peasant . . ."[9]

It is clear that Black recognised the risk to the child of being nursed by a tuberculous woman, more than 100 years before it became generally accepted in Northern Europe that tuberculosis is an infectious disease. Despite the obvious problem in medical ethics, Black had apparently no doubt about what was the proper course of action to take.

Whilst Scott's account left one with the feeling that Black must have been a remarkable physician, further evidence was needed to confirm this impression. Initial prospects seemed bleak as no evidence could be found of Black having preserved any case records among his papers and a search of the medical literature revealed only one report on a patient who had been under his care.[10] However this solitary case did provide a clue to another source of information as it described a stroke with an unusual course in Adam Ferguson, the celebrated moral philosopher of the Scottish Enlightenment. In directing attention to the other great figures of the Enlightenment it was readily established that several of them had been patients as well as friends of Black, including the illustrious David Hume.

A surprisingly detailed picture emerged of Black's management of Hume in his terminal illness from a study of the philosopher's correspondence and the letters of his literary friends and physician. Hume was acutely aware of the dangers of contemporary 18th century medicine and had once written to a sick friend:

"I entreat you, if you tender your own Health or give any Attention to the Entreaties of those that love you, to pay no regard to Physicians . . .   You cannot pay a moderate Regard to them: Your only Safety is in neglecting them altogether."[11]

Despite these views, Hume consulted Black as a patient in March 1776, having formed the opinion, over a number of years, that he was a man of sound judgement. Hume complained of recurrent diarrhoea, colicky abdominal pain, a sensation of excessive body warmth at night, physical weakness and progressive loss of weight for approximately one year; there were also indications of intestinal bleeding.[12] Cancer of the large intestine is the most probable cause of these features in a 64 year-old man and the much rarer possibilities, such as severe ulcerative colitis or regional enteritis, would carry an equally hopeless prognosis in the 18th century.

Black had no doubt about the presence of incurable progressive disease and also knew that his patient strongly suspected that his illness was lethal, Hume's mother having

died at his age of similar trouble. He decided that the philosopher's peace of mind would be best served by informing him of the true prognosis.[13] This action was greatly appreciated by Hume and it enabled him to put his affairs in order, write a brief autobiography and make gracious farewells to his numerous friends.

Hume readily accepted Black's advice to spend his remaining days in the comfort of his own home in St. David's Street, where he could enjoy his books and the pleasure of his friends. It was not long, however, before this plan of management was disrupted by the well-meaning, but wholly misguided, actions of Hume's distinguished friends, who were quite unable to accept that nothing could be done to save him. They tried to persuade Hume to travel to London to see Sir John Pringle, an eminent physician who, in his earlier days, had been professor of moral philosophy in Edinburgh. Although he initially refused to accept this advice he later changed his mind in order to please his friends. Black's unsuccessful attempt to dissuade him from setting out on this mad excursion ended in Hume teasing his physician. "Have you no reason against it", he said, "but an apprehension that it may make me die sooner? — that is not reason at all."[14]

Hume left for London at the end of April and, after a short stay in the metropolis and a longer period in Bath, he managed to reach home at the beginning of July, in a "very shattered condition", the jolting of the carriage on his emaciated frame having caused considerable pain.[15] During his visit to England, Hume had been seen by three doctors, each of whom had given a different opinion. Sir John Pringle had declared that he saw nothing alarming in Hume's case [16] and that he had a stricture of the colon which would be easy to cure.[17] He referred him for treatment to Dr. Gusthard, in Bath, and it was shortly after he arrived there that Hume wrote to his friend Hugh Blair:

"He [Dr. Gusthard] assures me that my Case is the most common of all Bath Cases, to wit, a bilious Complaint, which the Waters scarcely ever fail of curing; and he has never had a Patient of whose Recovery he had better hopes. Indeed the Waters in the short trial I have made of them (for I have been here only four days) seem to agree very well with me; and two days ago I found myself so well, that for the first time I began to entertain hopes of a Reprieve."[17]

These hopes were soon dashed by an exacerbation of symptoms, and arrangements were made for Hume to be seen by the famous anatomist and surgeon, John Hunter, who happened to be in Bath at the time. Hunter felt a tumour in his liver and was of the opinion that this was the seat of all his troubles.[18]

A few days after Hume's return to Edinburgh, Black summoned two colleagues to see his patient, Francis Home, President of the Royal College of Physicians and William Cullen, professor of the practice of physic. According to Hume this "grand Jury of Physicians" were all agreed that the opinions expressed by the doctors in England were absurd and erroneous. Whilst they confirmed Hunter's

Date (1776)	Medical attendant	Diagnosis	Prognosis (given to patient)
March	Joseph Black (Edinburgh)	Intestinal haemorrhage with colic and diarrhoea	fatal
May	Sir John Pringle (London)	Intestinal stricture	excellent
May	John Gusthard (Bath)	A common bilious complaint	excellent
June	John Hunter (visiting Bath)	Liver tumour	not divulged
July	Joseph Black, William Cullen & Francis Home (Edinburgh)	Intestinal haemorrhage and liver tumour	fatal

Table 4. David Hume's Terminal Illness

finding of a tumour in his liver, they said that is was so small as to be of little importance in comparison with the haemorrhage he was experiencing from the bowel.[19]

Hume died in August 1776 and in his last letter to his London publisher he wrote:

"This, Dear Sir, is the last Correction I shall probably trouble you with: For Dr Black has promised me, that all shall be over with me in a very little time: This Promise he makes by his Power of Prediction, not that of Prescription. And indeed I consider it as good News: For of late, within these few weeks, my Infirmities have so multiplyed, that Life has become rather a Burthen to me. Adieu, then, my good and old Friend
David Hume"[20]

The 'diagnoses' which were made during Hume's illness are shown in Table 4 and although there is no evidence of a post-mortem examination having been performed, three are compatible with cancer of the large intestine with secondary tumour spread to the liver, the exception being a "common bilious complaint"; however, such a conclusion was hardly possible in the 18th century as knowledge of the mechanisms of tumour spread was in its infancy. When we look at the differing prognoses (Table 4), it is obvious that Black and his Edinburgh colleagues had recognised a clinical pattern of disease associated with a fatal outcome. Although John Hunter did not inform Hume of his prognosis[18], it is almost certain that he regarded it as poor. From this comparison of contemporary medical opinion on Hume's illness it is evident that Black displayed sound clinical judgement. It is also apparent from the enlightened way in which he managed his dying patient that he had accurately assessed Hume as a man, this view being in accord with the opinion of Black's contemporary, John Robison, who wrote:

"... were I to say what intellectual talent Dr Black possessed in most uncommon degree, I think I should say it was his judgement of human character..."[21]

It may be asked how Black managed the everyday problems of medical practice, such as anxiety and depression. As his close friend James Watt suffered from these symptoms we gain some insight into Black's methods from the letters they exchanged. Depending on what he considered to be the underlying cause Black would offer practical assistance, sympathy or friendly advice. Watt was, in the modern jargon, a 'workaholic', and advice was frequently given on the need to take time off work and enjoy some social life. On one occasion he advised Watt to follow his example in buying a horse and riding as much as possible, provided, he cautioned, "you give your horse no corn for that makes them intolerably foolish and willfull".[22]

In concluding this assessment of Black as a physician, his performance in the management of his own illness merits consideration, surely a difficult test by any standard. In the Spring of 1797, when Black was 69, he began to experience "rheumatism and weakness" in his limbs. He had difficulty in walking and in climbing stairs and was unable to bear the jolting of a carriage.[23, 24] By the following winter his symptoms had become more severe:

"... I cannot move without a trusty stick to prop me in each hand while I drag one foot after the other. My figure and motions are too bad to be seen on the street."[25]

His previous medical history included recurrent febrile episodes dating from youth; these were accompanied by cough, expectoration and the spitting of blood[26, 27] and were almost certainly attributable to bronchiectasis, a disease in which the larger airways in the lungs are permanently damaged, usually as a result of infection in childhood.

For many years before the onset of his rheumatic symptoms Black had been taking a vegetarian diet, his only source of animal protein being milk. Since Edinburgh

Figure 4. The Meadows: scene of Joseph Black's "fresh air and grass milk" treatment.

milk, in the late 18th century, was obtained from cows which spent their entire lives in dark, dirty, city byres, his diet was grossly deficient in vitamin D. In a letter to James Watt he explained that "The insipid diet is necessary for me on account of my lungs and it is fortunate that my stomach can bear it. I do feel however sometimes the want of something more stimulating . . ."[28] There is little doubt that Black was mistaken in his belief about the need for his dietary restriction.

His exposure to sunlight was also deficient. This stemmed mainly from a leg injury sustained about six years prior to the onset of his rheumatic symptoms[29], which prevented his customary out-door activities of walking and horse riding. When he did venture out he used a sedan chair[30], and after the onset of his 'rheumatism' he was too ashamed of his appearance to be seen on the street.

Treatment of rheumatic complaints at the end of the 18th century consisted of free blood letting, brisk purging, vomits of tartar emetic and mercury administration in the form of calomel. It is little wonder that Black decided to manage his own complaint.

After a year of suffering and disability he wrote to James Watt:

"I have taken a house on the south side of the meadow where I shall have ample accomodation and hope that the fresh air and grass milk will be of service to me."[28] In June 1798 he informed his nephew that he had just moved to this house in the country and that "It gives me the advantage of fresh and sweet air and the opportunity to saunter out and creep about the doors or into the little Garden which belongs to it." [25]

Meteorological observations recorded within a mile of Black's country residence reveal that the summer weather of 1798 was mostly fine.[31] Although the milk from cattle at grass during this season would provide the ailing physician with some vitamin D, the quantity is likely to have been small in relation to the amount he would himself produce in response to fresh air treatment[32], vitamin D or cholecalciferol being synthesised in the skin by the action of ultraviolet light on its precursor, 7-dehydrocholesterol.

Black's health improved dramatically, much to the delight of his friends, on his "fresh air and grass milk" treatment and by the end of the summer he was able to write to his nephew: "I have continued to grow better and stronger and lately made out a walk to my house in town and after a rest there, I walked back again . . .   I am now

40

quite free of rheumatic pains . . ."[29] On returning to his town house for the winter he was pleased to report: "I can easily go up and down stairs which I was scarcely able for last winter."[33] His progress was maintained and after he had enjoyed another summer by the meadow he died suddenly in December 1799 at the age of 71, the circumstances[34] suggesting that he had succumbed to a heart attack.

It remains to be decided whether Joseph Black's disabling rheumatic complaint was due to osteomalacia, the musculo-skeletal disorder which results from vitamin D deficiency in adults, or polymyalgia rheumatica, a disease of unknown cause whose clinical presentation may closely resemble that of osteomalacia.

Although spontaneous cure is usual in polymyalgia rheumatica, the average duration of symptoms, in the absence of steroid treatment, is three years[35] whereas in Black's case he was pain-free in 18 months. Sleep disturbance due to pain is common in polymyalgia rheumatica[35] and uncommon in osteomalacia and was not a feature of Black's illness.[24] Lack of exposure to ultraviolet radiation, deficient intake of vitamin D and a vegetarian diet are recognised factors in the causation of osteomalacia[36] and all were operating in Black's case for several years before he became ill. The marked improvement which followed the adoption of his "fresh air and grass milk" regime is the response which occurs in osteomalacia, there being little, if any, effect in polymyalgia rheumatica. From these considerations there would seem little doubt that Joseph Black had osteomalacia and that he achieved healing of his disease when cause and cure were both unknown. He was indeed a remarkable physician.

NOTES AND REFERENCES

1. John Robison 'Editor's Preface' in Joseph Black *Lectures on the Elements of Chemistry* (Edinburgh 1803) I, xxiv.

2. Douglas McKie (ed.) *Notes from Doctor Black's Lectures on Chemistry 1767/8* (Wilmslow, Cheshire 1966) 172.

3. D.L. Cowen 'The Edinburgh Pharmacopoeia' in R.G.W. Anderson and A.D.C. Simpson (eds.) *The Early Years of the Edinburgh Medical School* (Edinburgh 1976) 25.

4. Robert Christison 'On the Poisonous Properties of Hemlock, and its Alkaloid Conia' *Transactions of the Royal Society of Edinburgh 13* 383 (1836).

5. Royal College of Physicians of Edinburgh MS Ab 4 Sir John Pringle 'Medical Annotations' volume V, 80, extract of letter from Black to Pringle, 14 October 1773.

6. Léo Marion 'The Pyridine Alkaloids' in R.H.F. Manske and H.L. Holmes (eds.) *The Alkaloids, Chemistry and Physiology* I (New York 1950) 165.

7. Eric Robinson and Douglas McKie *Partners in Science* (London 1970) 208, letter 149 (Black to Watt, 9 September 1794), and 209, letter 150 (Black to Watt, 28 October 1794).

8. James Gray *History of the Royal Medical Society 1737-1937* ed. Douglas Guthrie (Edinburgh 1952).

9. J.G. Lockhart *Memoirs of the Life of Sir Walter Scott, Bart.* (Edinburgh 1837) I, 14.

10. A.J.G. Marcet 'Case of Professor Ferguson, drawn up by Dr Black, in May, 1797' *Medico-Chirurgical Transactions, London 7* 230 (1816).

11. R. Klibansky and E.C. Mossner *New Letters of David Hume* (Oxford 1954) 173, letter 90 (Hume to John Crawford, 20 July 1767).

12. E.C. Mossner and I.S. Ross *The Correspondence of Adam Smith* (Oxford 1977) 190, letter 152 (Black to Smith, April 1776).

13. J.Y.T. Greig *The Letters of David Hume* (Oxford 1932) II, 314, letter 519 (Hume to John Home, 12 April 1776).

14. Henry Mackenzie *An Account of the Life and Writings of John Home, Esq. Appendix* (Edinburgh 1822) 167-170.

15. Greig *op.cit.* (13), II, 329, letter 531 (Hume to Strahan, 27 July 1776).

16. *Ibid.* 315, letter 521 (Hume to Strahan, 2 May 1776).

17. *Ibid.* 315, letter 524 (Hume to Blair, 13 May 1776).

18. *Ibid.* 324, letter 526 (Hume to John Home, 10 June 1776).

19. *Ibid.* 328, letter 530 (Hume to John Home, 9 July 1776).

20. *Ibid.* 331, letter 534 (Hume to Strahan, 12 August 1776).

21. Robison *op. cit.* (1), I, xlviii.

22. Robinson and McKie *op. cit.* (7), 178, letter 127 (Black to Watt, 2 July 1789).

23. *Ibid.* 276, letter 184 (Black to Watt, 13 June 1797).

24. *Ibid.* 281, letter 186 (Black to Watt, 11 November 1797).

25. Edinburgh University Library MS Gen 874 volume V, f97, letter from Joseph Black to George Black, 6 June 1798.

26. Adam Ferguson 'Minutes of the Life and Character of Joseph Black, MD' *Transactions of the Royal Society of Edinburgh 5* (Part 3) 101 (1805).

27. Robinson and McKie *op. cit.* (7), 139, letter 98 (Black to Watt, 28 May 1784).

28. *Ibid.* 289, letter 191 (Black to Watt, 8 April 1798).

29. Edinburgh University Library MS Gen 874 volume V, f 104, letter from Joseph Black to George Black, 25 September 1798.

30. Robinson and McKie *op. cit.* (7), 186, letter 133 (Black to Watt, 1 December 1791).

31. John Playfair 'Meteorological Abstract for the years 1797, 1798 and 1799' *Transactions of the Royal Society of Edinburgh 5* (Part 2) 193 (1805).

32. T.C.B. Stamp, J.G. Haddad and C.A. Twigg 'Comparison of Oral 25-Hydroxycholecalciferol, Vitamin D, and Ultraviolet Light as Determinants of Circulating 25-Hydroxyvitamin D' *Lancet 1* 1341 (1977).

33. Robinson and McKie *op. cit.* (7), 302, letter 201 (Black to Mrs Watt, 1 February 1799).

34. *Ibid.* 317, letter 211 (Robison to Watt, 11 December 1799), and 320, letter 213 (Robison to Watt, 18 December 1799).

35. Ian Gordon 'Polymyalgia Rheumatica' *Quarterly Journal of Medicine 29* 473 (1960).

36. B.E.C. Nordin *Metabolic Bone and Stone Disease* (Edinburgh and London 1973) 71.

Black, Hope and Lavoisier
W P Doyle

When Joseph Black took up the Edinburgh chair in
1766 at the age of thirty-eight his researches virtually came
to an end, and henceforth his contribution consisted solely
in his teaching. Black, like Cullen before him, lectured in
English not Latin. According to his biographer[1] "His
voice in lecturing was low, but fine; and his articulation so
distinct that he was perfectly well heard by an audience
consisting of several hundreds." Black, an expert manipu-
lator, illustrated his lectures with experiments which were
simple, neat and elegant and judiciously contrived to
establish clearly the point in question without over-
elaboration. His lecture table was as spotless at the end of
a lecture as when he began. His lectures aroused so much
interest that they attracted intelligent outsiders who wished
to hear about Black's own discoveries and to learn something of the general nature of chemistry. The average
number of students attending his lectures in the 1790s was
225, and during his tenure of the Edinburgh chair he
must have taught chemistry to about 5,000 students and
thus was immensely influential in the diffusion of chemical
knowledge during the latter half of the eighteenth century.
It is in this period that the foundations of modern
chemistry were laid, the so-called Chemical Revolution, the
chief architect of which was the Frenchman Antoine
Laurent Lavoisier, born in 1743 and guillotined in 1794.

What I want to discuss is how Lavoisier's work gradually
became incorporated into Black's teaching and what is not
the same thing to what extent Lavoisier's views became
accepted by Black.

From the beginning of the eighteenth century chemistry
had been dominated by the theory of phlogiston originated
by Johann Becher and developed and popularised by
Georg Stahl. Briefly, the theory assumed that all
inflammable bodies and metals contained a common
principle, which was termed phlogiston from the Greek
phlox meaning flame. When inflammable bodies are burnt,
or metals are calcined, phlogiston escapes. When a metallic
calx is converted back to the original metal by heating with
e.g. charcoal, phlogiston is transferred from the charcoal
to the metallic calx which thus becomes the original metal.
During this period, although it was well known that metals
increase in weight on calcination while during calcination
phlogiston was supposed to be lost, this fact was often
ignored as of little importance. One explanation advanced
was that phlogiston had negative weight and hence when
it escaped from a metal during calcination an increase in
weight resulted. Practically every chemist adopted the
theory of phlogiston during the eighteenth century.

On the basis of work begun in 1772 and substantially
completed in 1777, Lavoisier formulated a quite different
theory of the nature of combustion, which may be summed
up in the four following statements:
1) Substances burn only in "pure air".
2) Non-metals such as sulphur produce acids on
combustion; hence the "pure air" was termed oxygen
(derived from Greek words meaning acid producer).
3) Metals produce calces on absorption of oxygen.
4) Combustion is not due to an escape of phlogiston but
to chemical combination of the combustible substance with
oxygen.

This theory was at first put forward as an alternative to
the phlogiston theory, but when Lavoisier had accumulated
sufficient experimental evidence he felt able to assert in
1783 that the phlogiston theory was not only unnecessary,
since all the experiments could be explained just as well on
the new theory, but was actually incorrect since some of
its consequences conflicted with experiment.

For our purposes, important sources of evidence are
the notes of Black's lectures as taken down by students.
During Black's lifetime many manuscript versions of his
lectures were made by students, other members of his
audience, and not infrequently by professional copyists.
They were often beautifully written, bound in several
volumes and commanded a ready sale.

From student notebooks it is clear that Black, like
nearly all his contemporaries, fully accepted the theory
of phlogiston. A typical presentation of his views is given
in notes taken from his lecture on inflammation in the
1782-3 session.[2] After expounding the theory he concludes
"The other Facts on which this opinion is founded are
no doubt very strong proofs of it". He then refers to the
conflict between the theory and the fact that the weight
after inflammation is greater than the original weight. He
explains that there is no conflict if phlogiston has negative
weight, argues at length that this is not an unreasonable
concept, and concludes, in what reads like an impassioned
plea to convince his audience:
"there is then no absurdity in supposing that the
Principle of Inflammability is exempt from the Laws of
Gravitation; Nay, we may even suppose, that it may
have a Tendency rather from the Center, than towards
it, so may have the effect of rendering a Body, even
lighter than it is without it. In gen! I may observe that
there are a great many Facts wh when they come to be
known are not credited. Our Notions of Credibility

depend very much upon what we have been accustomed to.

And as this is the only Case in w$^{\underline{h}}$ the addition of an Ingred$^{\underline{t}}$ diminishes the absolute weight of the Comp$^{\underline{d}}$ it appears incredible. But after you have seen some of the expm$^{\underline{ts}}$ to w$^{\underline{h}}$ I refer, you will be satisfied of the existence of this common Principle of Infl$^{\underline{y}}$ notwithstanding this seeming Difficulty."

Two men who appear later as key characters are Sir James Hall and Thomas Charles Hope. James Hall was primarily a distinguished geologist with an intense interest in chemistry. He attended Black's lectures in the session 1781-2. A letter[3] written to his uncle, William Hall, in 1781, gives a picture of him at University.

"We came to Edinburgh the next day — ever since that we have been hard at work with two lectures every day but Saturday — Natural Philosophy and Chemistry. The Chemistry is very entertaining and Dr. Black delivers his lectures in the best manner so that not a word is lost."

Thomas Charles Hope was the son of John Hope, professor of botany at Edinburgh. He studied medicine in the University of Edinburgh, attending Black's lectures, and graduated M.D. in 1787. In the same year he was appointed to the lectureship in chemistry in Glasgow where he remained until he returned to Edinburgh in 1795 as conjoint professor with Black. Hope attended Black's course in the 1781-2 session and re-attended in both the two succeeding sessions. His notes[4] of Black's lectures taken during the 1782-3 session survive and Hope describes them as "pretty Exact". They are most interesting because they were written on the right hand side of the page only and in the succeeding session he made additions and corrections on the left hand side of the page. The notes of 1782-3 on the theory of phlogiston are identical in substance to the typical presentation previously given. However, among the additions of the next session, 1783-4, we find notes indicating that Black was now incorporating some of Lavoisier's work into his lectures. An example, from the section on inflammation, is "Some Chemists in France suppose that no bodies contain the phlogiston. Dr. B. supposes that bodies contain the phlogiston". Another example, from the section on calcination of metals, is "Lavoisier supposes that calcination depends on the absorption of Empyreal air or = his acid principle and by this he accounts for the addition of weight and for the extrication of air from the calces".

Returning now to Sir James Hall, in the summer of 1783 he embarked on an extensive grand tour of Europe and he did not return to Scotland until the summer of 1786. Early in 1786 he was in Paris and there he met Lavoisier. He wrote[3] as follows to his uncle:

"I have only time to say that I am at last compleatly come over to your opinion. I am personally acquainted with Mr. Lavoisier & have received the greatest civilities from him — I have a standing invitation to dine with him every Monday — he has one of the clearest heads I ever met with & he writes admirably.

I have read most of his papers in the memoires of the Accademy. They begin in that for the year 1776 & continue till 1782 which is the last published the principal ones are the 1777 The [word crossed through] exclusion of Phlogiston is of this last date so you see it is no novelty — he spoke of this to me himself, he observed he said [que] ces idees commencent à germer and that they are talked of as new discoveries whereas they are of ten years standing at least".

A comment in Hall's last letter[3] from Paris is of considerable interest:

"Some time ago that a new foreign member was to be chosen, the accademy was divided between Black & Priestly the preference was given to the last for what reason I don't know I believe because he had published much. But all the chemists were clear for Black . . . "

Early in 1788 a full-scale debate on the new versus the old chemistry took place in the Royal Society of Edinburgh, Hall expounding Lavoisier's views and James Hutton defending phlogiston. The only information we have are the brief details as given in the Transactions of the Royal Society of Edinburgh.[5] We do not know whether or not Black was present although it is difficult to conceive that he did not attend some of the sessions but at the very least much of what was discussed must have been conveyed to him by his intimate friend Hutton.

One way in which Hall was influential in diffusing Lavoisier's views is that he and Hope were friends, a friendship perhaps dating from their days as fellow-students at Edinburgh, and after Hall's return to Scotland they had long discussions on the work of Lavoisier, and Hope became a convert to the new ideas. In the session 1787-8 Hope taught them to his class in Glasgow, thus becoming the first teacher in Britain to substitute the views of Lavoisier for the phlogiston theory. Hope spent the summer of 1788 in Paris and saw much of Lavoisier.

We come now to the years 1789 and 1790 in which Black and Lavoisier corresponded with one another. There are only 5 letters, 3 of Lavoisier's, 2 of Black's, but of such interest as to be well worth a paper in themselves. Here I deal with them briefly in their relevance to my central theme. In 1789 Black was elected a Foreign Member of the Royal Academy of Sciences of Paris; Lavoisier's first letter[6] to Black, of September 19, 1789, uses this to introduce himself:

"Sir — It is a member of the Royal Academy of Sciences of Paris who writes to you with the title of Confrère; it is one of the most zealous admirers of the depth of your genius and of the important revolutions which your discoveries have caused in the Sciences".
The letter then proceeds to its purpose — the introduction of its bearer, an emigré who is going to finish his education in Edinburgh; the bearer is bringing Black a copy of a work which Lavoisier has just published:
"you will find there a part of the ideas of which you have laid the first nucleus; if you have the goodness to give some moments to reading it, you will find there

the development of a new doctrine which I believe simpler and more in accord with the facts than that of Phlogiston. It is only with trembling that I submit it to the first of my judges and to him to whose support I most aspire."

If Black answered this letter we have no record of his reply.

Lavoisier's second letter,[7] of July 24, 1790 begins: "Sir, — I learn with an inexpressible joy that you are very willing to attach some merit to the ideas that I have been the first to profess against the doctrine of phlogiston. More confident in your ideas than in my own, accustomed to look upon you as my master, I should distrust myself as far as I have deviated without your approval from the road you have so gloriously followed. Your approbation Sir, dissipates my anxieties and gives me new courage."

The letter then goes on to introduce another emigré who is going to study in Edinburgh, and concludes:

"I will not be content until circumstances allow me to be going to present to you myself the proof of my admiration and to range myself in the number of your disciples. The revolution which is effected in France must naturally make useless some of those attached to the old administration, it is possible that I enjoy more liberty and the first use that I shall make of it will be to travel and to travel especially in England and to Edinburgh to see you there, to hear you there and to profit from your insights and from your advice."

Black's reply[8], dated October 24, 1790, begins by commenting favourably on the emigré student, and continues:

"You have been informed that I endeavour in my Courses to make my Pupils understand the new principles & explanations of the Science of Chemistry which you have so happily invented and that I begin to recommend them as more simple & plain and better supported by Facts than the former system, and how could I do otherways? . . .
The system you have founded is so simple & intelligible that it must be approved more & more every day and will even be adopted by many of those Chemists who have long been habituated to the former System: To gain them all is not to be expected, you know too well the power of habit which enslaves the minds of the bulk of mankind and makes them beleive & reverence the greatest absurditys. I must confess that I felt the power of it myself, having been habituated 30 years to beleive & teach the doctrine of Phlogiston as formerly understood. I felt much aversion to the new system which represented as an absurdity what I had believed to be sound doctrine this aversion however which proceeded from the power of habit alone has gradually subsided, being over come by the clearness of your demonstrations & consistency of your Plan and tho there are still a few particulars which appear to be difficultys, I am satisfyed that it is infinitely better supported than the former Doctrine; . . . But tho the power of habit may prevent many of the older Chemists from approving of your Ideas, the younger ones will not

be influenced by the same power; they will universally range themselves on your side of which we have experience in this university where the students enjoy the most perfect liberty of chuseing their philosophical opinions. They in general embrace your system and begin to make use of the new nomenclature in proof of which I send you two of their inaugural dissertations in which chemical subjects were chosen; these Dissertations are wrote entirely by the students; the professors have no share in them. We read them before they are printed to see that there are no gross absurditys in them & give our advice if any are found".

Lavoisier replied,[9] on November 19, 1790, that he could not have received a more agreeable present than Black's letter. He goes on: "I have another favour to ask you, but on which I ought to await your consent; it is to wish to allow me to publish the translation of it in the Annales de Chimie." Black's reply[8], of December 28, 1790, consents to Lavoisier's request in the following words:

"It gave me pleasure also to find that you are satisfyed with the avowal I have made of my approbation of your System of Chemistry. You have my full consent to publish my letter. This consent I consider as a tribute I owe to truth and the eminent Rank you hold as a promotor & Patron of the Science of Chemistry".

Black's letter approving Lavoisier's system was translated into French and published in the *Annales de Chimie* in March 1791.

While reading this fascinating correspondence I was interested to see if it were possible to identify the two dissertations which Black sent to Lavoisier as evidence that the students embraced Lavoisier's system. Examination of the Edinburgh M.D. graduation theses showed that there are only two, both written in 1790, which fit the bill — one by John Gibney[10] and the other by Alexander Anderson.[11] Both contain numerous references to Lavoisier's work. One of the two (Anderson's) contains a less than diplomatic error — you may remember that Black told Lavoisier that the professors read the theses before they were printed to see that there were no gross absurdities in them. However in this thesis 'Antoine Laurent Lavoisier' appears as 'John Henry Lavoisier'.

That Black had by this period definitely accepted Lavoisier's views is further evidenced by a letter[12] from him to a friend in July 1792 in which he described how "for my part I now, tho I had a reluctance at first, find the french theory so easy and applicable that I mostly make use of it".

We return now to Thomas Charles Hope. While at Glasgow, Hope made his mark by his research on strontia, and by the popularity of his lectures. This promising beginning gave Black, then in declining health, the idea of having Hope as his assistant and subsequent successor. He accordingly approached Hope in 1795, obtained the agreement of the Edinburgh Town Council, and on November 4th the Council appointed Hope conjoint professor with Black. In the session 1795/6 Hope delivered only a few lectures.

The session 1796/7 was the last session in which Black lectured. In the Edinburgh University Library there is a manuscript copy of notes[13] from the lectures delivered by Black and Hope in the session 1796/7. Black and Hope alternated, each doing six blocks, although these were not of uniform length. It was Black, and not Hope, who dealt both with inflammation and with the calcination of metals, and who treated them in terms of Lavoisier's views. The section of these notes referring to the calcination of metals contains the last public exposition by Black concerning the old and the new chemistry:

"The Stahlians accounted for it on the same principles as for combustion in general. That by calcination they lost their Phlogiston, and recovered it by reduction, which so far was certainly intelligible, but there was one very great objection, that the calces which had lost something were always heavier than the metals originally. To explain this they framed hypotheses altogether unintelligible, one of which was that Phlogiston had a principle of levity. . . .
Mr. Lavoisier therefore and his friends have formed a thory quite opposite to that of the Stahlians. They conclude metals are as simple bodies as any we are acquainted with, that the calx is a union of the metal and oxygen. That some calces retain oxygen so feebly that it can be reduced by heat alone. That those which retain it strongly, are reduced by the action of inflammable bodies: if with charcoal, carbonic acid is formed, if with hydrogen, water. All this is proved by incontestible experiments. The only circumstances the French Chemists do not satisfactorily explain is the production of heat and light."

A man's epitaph is the opinion of those who knew him. Black died on December 6, 1799, and I wish to conclude by quoting from the minutes of the Senatus of the University of Edinburgh, asking you to bear in mind that Senatus minutes of that period were normally *very* brief, and that in 1799 Black would be known only by reputation to most of the students.

At the Senatus meeting of December 9th it was recorded[14] that:

"In consequence of the death of Dr. Joseph Black, which happened on friday last, the 6th Inst. the Senatus Academicus, as a mark of respect to the memory of their late eminent colleague (upon the report of the Principal that it would be most acceptable to the relatives) resolved to accompany the funeral in their gowns, preceded by the mace, on friday next at one o'clock; & understanding that many students wished to testify a similar respect, they agreed that intimation should be made in the classes desiring all such Gentlemen to meet with the Principal and Professors in the Library, at the proper hour, that they might proceed altogether in one Academical body."

The Senatus Minutes of December 13th 1799 record:

"In consequence of their former appointment, they went in procession in their Gowns, preceded by the Mace, and followed by a numerous body of students (who had assembled in the great room of the Library) to the house of the late Dr. Black in Nicholson's Street, & from thence accompanied the funeral, by Chapel Street to the Greyfriars Church-yard. After the internment they returned to the College, through Brown's Square & Argyll's Square."

NOTES AND REFERENCES

1. Joseph Black *Lectures on the Elements of Chemistry, delivered in the University of Edinburgh; by the late Joseph Black, M.D.* (Edinburgh 1803) lxii.

2. Edinburgh University Library [E.U.L.] MSS DC. 2.41, p.140.

3. V.A. Eyles 'The Evolution of a Chemist, Sir James Hall' *Annals of Science 19* 153-182 (1963).

4. E.U.L. MSS Dc. 10. 9.

5. *Trans. Roy. Soc. Edin. 2* 26 (1790).

6. E.U.L. MSS Gen. 874/IV, ff. 41-2.

7. *Ibid.* ff. 43-4.

8. D. McKie 'Antoine Laurent Lavoisier, F.R.S., 1743-1794' *Notes and Records of the Royal Society of London 7* 9-13 (1949-50).

9. E.U.L. MSS Gen. 874/IV, ff. 45-6.

10. John Gibney *Disput . . . inaug. de aethere* (Edinburgh 1790).

11. Alexander Purcell Anderson *Tentamen . . . inaug. de compositione acidi sulphurici* (Edinburgh 1790).

12. E.U.L. MSS Gen. 873/III, f. 232.

13. E.U.L. MSS Gen. 48D.

14. Edin. Univ. MS Senate Minutes: E.U.L.

Joseph Black and John Robison
J R R Christie

"[Dr. Black] went into the Lavoisierian doctrines *tête basse*, more (I think) like a pupil than like a great Master . . . and I do not think he has done this neatly."[1]

"As a professor, and conscious of his own Rank in the opinion of the public, [Dr. Black] should have done somewhat more than merely acknowledge the Value of Lavoisier's Observations and Experiments. He should have stated the Objections which a Philosopher would still make."[2]

"[Dr. Black's] theory of Lime is tedious beyond bearing . . . the reader cannot but see the keeping up of the great discovery till the very last."[3]

"Dr. Black seems to have turned his whole attention to rendering his Lectures as popular and profitable as possible, by a neat exhibition of Experiments — he multiplied these, without any new Views."[4]

"I question whether [Dr. Black] would ever have put his doctrine of latent heat on a footing that would have given satisfaction to the public. He saw clearly the broad principle, and took no pains by measurable experiments to ascertain the equality of the absorbed and Emerging heats."[5]

"[Dr. Black's] notion of rival attractions, and of one attraction weakening another etc., etc. are not like those of a person accustomed to consider mechanical actions with that obstinate simplicity that is indispensably necessary for clear conceptions and accurate reasonings."[6]

The foregoing quotations are taken from letters written by John Robison in the period December 1799 to April 1803, the three years and four months during which he was engaged in the Herculean labour of preparing Joseph Black's lecture notes for the press. In summary form they tell us that John Robison thought Joseph Black had cravenly capitulated to Lavoisier's Chemical Revolution, and that this involved a dereliction of philosophical duty; that his lectures were old-fashioned and tedious; that his lectures were modified with a view to their popularity and profitability, at the expense of treating innovations in chemical science; that Black's prosecution of experimental enquiry was deficient; and that Black was a poor theoretician.

It is not my purpose to substantiate Robison's judgments. This would involve my flying in the face of the considered opinions of almost all historians and memorialists who have written on Black, and who have indeed testified to his brilliant experimental acumen, his inspirational pedagogy and his theoretical prowess. Neither, however, is it my purpose to take John Robison to task for his pronouncements, to convict him of gross misjudgments and sentence him; or else perhaps suspend conviction because his pronouncements were uttered while the balance of his mind was disturbed, for such in fact was the case, as we shall see.

Instead, I use these quotations from Robison in the first instance to ask a question. How was it that Robison came to hold such opinions, yet produced a published text, Black's *Lectures on the Elements of Chemistry*, where his derogatory views are only occasionally glimpsed, where instead the dominant picture of Black is the stereotype we all know, the meticulous inductive philosopher, the great original discoverer of latent and specific heat, the ingenious formulator of the fixed air theory of causticity? This, of course, is to raise a doubt — what sort of text is it which purveys the opposite of the author's beliefs? This doubt is greatly deepened and intensified as one reads through Robison's correspondence for the period in question. It provides a blow-by-blow account of Robison's editorship of Black's *Lectures,* and to my mind it demonstrates quite simply the fact that the two volumes of the *Lectures,* along with its substantial Preface and Notes, constitute a profoundly problematic document. This makes a basic strategic point for all Black researchers. Sooner rather than later, we all turn to Robison: for biographical detail, for material crucial in the reconstruction of Black's scientific researches, for Black's methodological views, for his broad conception of the frame of nature. Robison's edition of the *Lectures* is the Black researcher's main source. The letters which reveal its problematic nature have been available in published form for over a decade. My proposition is that it is high time we started to examine their implications for our main source; and in delineating this proposition I hope to persuade you that the *Lectures* cannot be taken on trust, that we need instead to manufacture a key which enables us to determine on what points and to what extent Robison is to be approached in a critical spirit, as opposed to an unquestioning one.

The only way to manufacture such a critical key is to provide a full account of Robison's editorship and the circumstances which surrounded it. This editorship was a long, complex and painful process, in both an intellectual and psychological sense. This is not the place to give a full-

H. Raeburn Esq.r pinx.t   C.Turner sculp.t

JOHN ROBISON, L.L.D.

Professor of Natural Philosophy in the University of Edinburgh &c.a &c.a

Edinburgh Published by the Proprietor Feb.y 27 1805.

Figure 5. John Robison: mezzotint by Charles Turner after Sir Henry Raeburn, 1805.
(Reproduced by permission of the Scottish National Portrait Gallery)

blown narrative, but I can, I hope, provide a sense of the elements involved, and the way in which they hang together.

Firstly, then, to the personal aspects of the case. Robison was sixty years of age in 1799 when he accepted what he called "this last Office of friendship" and undertook to edit Black's lectures.[7] By this time he was a widely-travelled, highly-experienced and highly-respected man of science. He had been born the son of a Glasgow merchant, and was educated at school and university in Glasgow, where he received his M.A. in 1756. Before travelling to London in 1758 he had made the acquaintance of Black, then in the early years of his Glasgow lectureship in chemistry. In 1759 Robison sailed with Wolfe to Quebec, as tutor to the son of Admiral Knowles, and on his return in 1762 was appointed by the Board of Longitude to assist in the marine trials of Harrison's chronometer. In 1766, with Black's support, he succeeded Black in the Glasgow chemistry lectureship, but by 1770 was travelling once more, to Russia with Admiral Knowles, where in 1772 he became professor of mathematics at the Cronstadt Academy for the Imperial Sea Cadet Corps. He returned to Scotland in 1774, enticed back by Black's and Principal William Robertson's offer of the chair of natural philosophy at Edinburgh, a position he held until his death in 1805. Robison was also general secretary to the Royal Society of Edinburgh from its inception in 1783 until a few years before his death. His abilities as a man of science were in my view very considerable, particularly in the field of pure and applied mechanics, though historians of physics have tended to neglect him. He is credited with being the first man to think of applying steam-power to locomotion.[8] James Watt thought Robison the "man of the clearest head and most science I have known", and Watt spoke as one acquainted with most of the scientific luminaries of his age.[9]

This extensive and varied background ought to have made Robison an ideal editor. Besides his native abilities, he had known Black in the 1750s and early '60s, the crucial creative years of Black's life. He had been a colleague of Black's on the professoriate at Edinburgh for a quarter of a century. Yet, in December 1799 Robison was a man afflicted. He had financial difficulties, and the complicated and quarrelsome financing of the editorship, which involved Black's relatives and Robison as principal parties, together with external adjudicators for copyright shares and a paid amanuensis, contributed substantially to the long-drawing out of the editing and the delaying of the completed work's final appearance.[10] Robison was also an ill man, suffering constant pain with an ulcerated stomach. This pain was itself distracting, but the distraction was considerably worsened by Robison's means to ease the pain. For this he took opium, a fact which increased his fascination for his students, but which also had deleterious effects upon his powers of concentration.[11] He admitted as much to James Watt, saying specifically that his opium therapy had clouded and confused his mind while he was editing Black's *Lectures.*[12] Further, and perhaps partially due to the opium therapy, Robison was gripped by lurid and delusional political fantasies of a paranoid nature. The finance, the

illness and the opium all naturally combined to prolong and exacerbate the process of editing. The political paranoia, I will shortly show, had even more serious effects.

These more serious effects, however, need to be understood against the intellectual considerations involved, which were of several sorts. First of all, Robison had no substantial text of Black's to work from. Black's lecture notes were very incomplete and piecemeal, generally unorganized, so Robison and his amanuensis, working from student lecture notes, were obliged to fill in considerable gaps.[13] There were also gaps to be filled in relating to the substantial experimental basis of Black's lectures. Black would apparently extemporize on numerous experiments performed in laboratory and in front of class, and their reconstruction cost Robison much time.[14] At the same time, Robison was forced to the unpalatable conclusion that Black's lectures simply were not very good. They were pitched at a very basic level, were dully repetitive, and relied on the entertaining visual appeal of the experiments, impeccably performed by Black.[15] Robison's first problem, of how to turn this distressing legacy into a chemical text-book, was overlain by another problem, equally distressing. Robison had certain expectations of what this posthumous work of Black's should show, and he believed that the educated public shared those expectations. Black for Robison was a great scientist, possessor of an ingenious and elegant philosophical mind.[16] There was no way in which Black's elementary and repetitive lectures could substantiate and conform to that image of the elegant and philosophical Black. Robison did try to evade this problem, by suggesting that he publish only Black's famous work on causticity and heat, but Black's relatives held him to the contract, which was for publication of the full body of lectures.[17] So Robison had to settle for a revised presentation of Black in the lectures themselves — Black the careful pedagogue, scrupulously leading his students through the basic aspects of chemical practice and theory. He used his Preface and Notes to vindicate Black's originality and philosophical commitment. However, there is no doubt that once Robison had partially discarded his preferred image of the philosophical Black under the impact of the awful lectures, this opened the flood-gates of doubt. The longer the editing went on, the more critical of Black Robison became. By the time of publication, Robison had reached those conclusions already outlined — that Black was a poor teacher, bent on pecuniary profit rather than educational clarity; that Black was an inadequate experimenter, failing to prosecute his work on heat with sufficient experimental energy and rigour; that Black was no theoretician, lacking the ability for close, concentrated thought necessary to the construction of acceptable theories; and lastly that Black was no philosopher.[18]

Black no philosopher — this was the most drastic of Robison's conclusions, and by it he meant that Black lacked a reasonably systematic, methodologically well-founded approach to chemistry. How had this come about, how had the brilliant and youthful Black who had made such a profound impact on Robison in the 1750s and 1760s, become unphilosophical? The answer, in a word, was Lavoisier. The

Lavoisierian revolution presented yet another series of difficulties for Robison the editor. It posed the profound problem of how to render what was basically a pre-Lavoisierian chemistry relevant to a post-Lavoisierian world of chemical theory and practice. Black had been unphilosophical in two senses for Robison. Firstly he had failed to incorporate Lavoisier's work *systematically* into his lectures, but instead had carried out a very bitty, piecemeal accommodation to Lavoisier's doctrines.[19] Secondly, and somewhat contradictorily, Black had failed wholly to take a scientifically-argued stand against Lavoisier, had failed, in Robison's words to "make those objections which a philosopher would still make".[20]

It was at this point in Robison's editing that his political paranoia took firm hold. A few years before, in 1797, Robison, acutely troubled by the political development of Revolutionary Europe, had published a work which was designed to show how the late eighteenth century political revolution was based in a conspiratorial movement originating in Free Masonry.[21] All the details of this conspiracy need not detain us now; but it should be pointed out how, for Robison, this movement attained its ends at the level of ideology. While overtly claiming to pursue economic progress and philosophical rationality, the conspirators had in fact promoted a diabolical triad of philosophical doctrine, consisting of scepticism, materialism and determinism. Scepticism, to undermine beneficial and valid belief in the foundations of knowledge, and from there, religion.[22] Materialism, to undermine distinctions between God, man and animal, thereby incorporating moral feeling and conduct within a monistic system which treated all mental and material phenomena as merely modifications of an aetherial fluid.[23] And universal determinism, to rid the universe of a commanding Deity who maintained and guaranteed the natural and moral order, including the free will of His created human beings.[24] In combination, scepticism, materialism and determinism promoted irreligion and subverted moral and political authority. The chief villains of this philosophical conspiracy were French, though Priestley too qualified, as a materialist. The villainous French were such figures as Turgot, Condorcet, Diderot, D'Alembert and Laplace — the leading scientific figures of the French Enlightenment and Revolution, men who had not only promoted such doctrine, but had even recruited heads of state to their programme — Catherine of Russia, Gustavus Adolphus of Sweden, Frederick of Prussia, all were involved, and now in Britain the conspiracy was growing apace.[25] Faced with this imminent Apocalypse, Robison's answer, in the philosophical sphere, was to reply with the names and work of the British inductive heroes, Bacon and Newton. Thus Newton's Providential God gave the lie to Laplace, and empirical inductivism was set against French deductive rationalism.[26]

Now how did all this affect editing Black's *Lectures*? Robison saw Lavoisier as one more element and variant in the French conspiracy. He and his followers had acted dishonestly toward Black, publishing his letters without permission in order to gain support for the new chemistry.[27] In general they had acted in an authoritarian manner, brooking no opposition to their doctrines, dissolving the memory of older chemical theories with their new nomenclature which itself incorporated the new concepts.[28] Finally Robison equated French chemistry with revolutionary Jacobinism. As in politics, so in science. Black, the former king of chemistry, had been deposed from the throne, and chemistry was now ruled by a Revolutionary Committee, a Committee whose scientific methodology structurally mirrored their authoritarian politics.[29] Instead of a critical and cautious empirical analysis, the oxygen theory of combustion and acidity was placed at the head of a synthetic (non-analytic) system, where chemical behaviour was deduced from the theoretical premises, and confirmed only by, to Robison, unpersuasive gravimetric arguments. This, said Robison, was "the ruin of philosophy".[30]

It was primarily to combat the pernicious French chemistry that Robison formulated the intellectual *leitmotif* of the edited *Lectures*. He simply swallowed his own advice to the British public in his counter-revolutionary tract, and saturated Black's *Lectures*, from the opening words of the dedication, through the long introductory preface, the text and the extensive notes to the text, with laudatory references to inductive methodology. This done in a variety of ways — straightforward prescription, biographical and conversational detail from Black in Preface and Notes, the presentation of material in the actual body of the text.[31]

Inductivism, from there on, served as the principle means to rescue Black's philosophical reputation. Robison used it to underpin what now became the cautious analytic presentation of the thorough pedagogue following, as Robison claimed, a strict Baconic ordering of observation and generalization.[32] Similarly, Lavoisier's empirical results could be shorn of their synthetic deductive framework, and incorporated within the inductive scheme.[33]

There remained for Robison one final and crucial threat to be eliminated. Black's reputation was of course founded on his original discoveries. In the case of fixed air and causticity Black's reputation was secure, for the work had been published early on, in 1756. However, Black had never published his work on heat, so his claims for priority in that case were at risk. Robison, his ire thoroughly aroused, his paranoia discerning enemies around every corner, saw a counter-claim to Black's priority in the work of the Swiss natural philosopher Jean Andre de Luc, who according to Adair Crawford, had observed latent heat in 1755 or 1756.[34] Unfortunately for de Luc, he had been involved in an abortive project to publish Black's work on heat in the early 1780s.[35] Robison concluded that de Luc had plagiarized Black's work, and on this plagiarization had been erected false claims for priority of discovery.[36] Robison therefore published in the *Lectures* a vituperative assault on de Luc's probity, together with extracts from Black's notebooks which showed that Black had discovered latent heat as early as the years 1754-6.[37]

With de Luc villified, French pretensions tamed and Black's philosophical reputation saved by induction, Robison could rest easy. The historian however, cannot.

Suppose one wanted to produce an account of Black's methodology. A prime piece of evidence would be conversation between Black and Robison which Robison printed in the *Lectures,* during the course of which Black spoke strongly against the deductivism of the French. The trouble with that conversation is that it can be reconstructed, virtually phrase for phrase, from Robison's own outbursts in his editorial correspondence.[38] Similarly, suppose you were to seek to show the relationship between Black's methodology and his work on heat — something which many writers on this area have done, concluding that Black's work avoided theoretical presuppositions on the nature of heat.[39] Should one not rather think of Robison's avowed hostility towards the explanation of field and dynamic phenomena by materialist fluid theories associated by Robison with materialism, and ask if Robison's cautious inductive Black was not more hypothetically inclined than Robison indicated. Suppose again, the basic task of even dating Black's discovery of latent heat accurately — does one follow McKie and Heathcote, and Robison's preface, and date the discovery according to its experimental validation in the late 1750s and early 1760s?[40] Or does one follow Robison's Notes (and H.T. Buckle) and date the discovery in the year 1754-6?[41] Only to realize of course that this evidence is printed in the context of refuting de Luc's claim for priority, and given Robison's state of mind and editorial licence, may be suspect.

In short, understanding Joseph Black first involves understanding John Robison and his editorship. Only when one has grasped Robison's intentions and strategy can one begin to use them as a key, and an imperfect key at that, to disentangle a putative historical reality from Robison's complex mythography.

On the basis of what I have said, I think it is possible to lay down some guiding principles for any historian utilizing Robison's edition of Black's lectures. Firstly, there are four important areas where Robison is to be treated with considerable caution. Black's reception of and attitude towards Lavoisier's chemical revolution; Black's methodological commitments; Black's theoretical commitments, especially where aethers/imponderable fluids are concerned; and Black's discovery of latent heat. On each of these topics Robison is in fact internally inconsistent in what he wrote, and the editorial correspondence reveals the particular pressures, hostility to the French, vindication of Black's originality, which produced his inconsistency. The practical corollary of this caution is that, wherever possible, Robison's accounts should be compared and correlated critically with any other collateral evidence. In the case of the lectures themselves we have reasonable security, for there are numerous copies of lecture notes by Black's students throughout his years in Edinburgh.[42] These have the added advantage of giving a developmental view of Black's opinions from 1766, as opposed to the final summary of the published edition. Elsewhere we are less fortunate. There are no lectures I know of from Black's Glasgow period — these may improve our knowledge of Black's work on heat should any come to light. There are, however, other resources for Black on heat, apart from

Robison's contradictory accounts — particularly Black's own recollections, his correspondence with Cullen, and excerpts from Black's notebooks published by Adam Ferguson in his biographical memoir on Black.[43] Thus, in student lecture notes, in letters, in other accounts of Black, we do have resources which can partially resolve the problems Robison created.

I should like to finish on a less austere note. While I am not willing to suggest that we adopt the dismal view of Black's abilities which my opening quotations from Robison provided, we may want to take some account of his opinions — if Black was unable to cope intellectually with Lavoisier's innovations, if Black's early work on heat reveals a speculatively probing and not a cautiously empirical mind, then we do no service to his memory by continuing to bow down before inductivist and positivist icons of the man whose birth we celebrate. We want surely to celebrate and understand Black's life as he lived it, with all that it entailed in terms of human effort and ambition, success and failure. Similarly we may celebrate John Robison. He undertook his work as a 'last office of friendship', and as an act of friendship, it has proved marvellously successful.

NOTES AND REFERENCES

1. Robison to Watt, July 23, 1800: E. Robinson and D. McKie (eds.) *Partners in Science: Letters of James Watt and Joseph Black* (Cambridge, Massachusetts 1970) 344.

2. Robison to Watt, October, 1800: *ibid.* 356.

3. Robison to Watt, July 23, 1800: *ibid.* 344.

4. Robison to Watt, February 25, 1800: *ibid.* 339.

5. *Ibid.*

6. Robison to Watt, October, 1800: *ibid.* 359.

7. Robison to Watt, December 18, 1799: *ibid.* 322. The biographical account of Robison which follows is taken from the *Dictionary of National Biography.*

8. L. Darmstaedter *Handbuch zur geschichte der Naturwissenschaften und der Technik* (Berlin 1908; Kraus reprint, 1960) 204.

9. Watt to Murihead, February 7, 1805: Robinson and McKie *op. cit.* (1), 389.

10. Robison to James Black, September 16, 1802: D. McKie and D. Kennedy 'Some letters of Joseph Black and Others' *Annals of Science 16* 161-2, 164-5 (1963).

11. Henry Cockburn *Memorials of his Time* (Edinburgh 1856) 56.

12. Robison to Watt, April 19, 1803: Robinson and McKie *op. cit.* (1), 376-7.

13. Robison to Watt, July 23, 1800: *ibid.* 342-3.

14. Robison to George Black *junior,* August 1, 1800: *ibid.* 348.

15. Robison to Watt, July 23, 1800: *ibid.* 343-4.

16. Robison to Watt, December 29, 1799: *ibid.* 324.

17. Robison to George Black *junior,* August 1, 1800: *ibid.* 350.

18. See references (1) to (6).

19. Robison to George Black *junior,* August 1, 1800: *ibid.* 349.

20. Robison to Watt, October, 1800: *ibid.* 356.

21. J. Robison *Proofs of a conspiracy against all the religions and governments of Europe, carried on in the secret meetings of Free Masons, Illuminati, and Reading Societies. Collected from good Authorities.* (2nd edition, London 1797).

22. *Ibid.* 428-9.

23. *Ibid.* 483.

24. *Ibid.* 231.

25. *Ibid.* 520-21 (footnote).

26. *Ibid.* 230-31, 531.

27. Robison to James Black, September 16, 1802: McKie and Kennedy *op. cit.* (10), 166.

28. Robison to Watt, September 9, 1800: Robinson and McKie *op. cit.* (1), 352.

29. Robison to George Black *junior,* August 1, 1800: *ibid.* 349. See also p.356, and McKie and Kennedy *op. cit.* (10), 168.

30. Robison to Watt, October 1800: Robinson and McKie *op. cit.* (1), 357. See also pp. 345-6, 352, 356-7.

31. Robison to James Black, September 16, 1802: McKie and Kennedy *op. cit.* (10), 168. See also Robinson and McKie *op. cit.* (1), 357.

32. Robison to Watt, October 1800: Robinson and McKie *op. cit.* (1), 357.

33. *Ibid.*, also p.345.

34. A. Crawford *Experiments and Observations on Animal Heat* (2nd edition, London 1788) 71-2.

35. Watt to Black, December 13, 1782: Robinson and McKie *op. cit.* (1), 117.

36. Robison to Watt, February 25, 1800: *ibid.* 336-7.

37. J. Robison (ed.) *Lectures on the Elements of Chemistry, by Joseph Black, M.D.* (Edinburgh 1803) I, 523-7.

38. *Ibid.* 547; compare with Robinson and McKie *op. cit.* (1), 345-6, 356-7.

39. E.g. D.McKie and N.H. de V. Heathcote, *The Discovery of Specific and Latent Heats* (London 1935) 27-8.

40. *Ibid.* 30-6.

41. H.T. Buckle *On Scotland and the Scotch Intellect* ed. H.J. Hanham (Chicago 1970) 312, footnote 149.

42. See the accompanying article in this volume by William Cole.

43. Adam Ferguson 'Minutes of the life and character of Joseph Black, M.D.' *Transactions of the Royal Society of Edinburgh 5* (Part III) 101-117 (1805).

Manuscripts of Joseph Black's Lectures on Chemistry
William A Cole

The following list of manuscripts of student lecture notes had its origin in connection with a project conceived many years ago, when a copy of Black's lectures was acquired by the present author. The date of the manuscript, 1788, and the references in it to Lavoisier, suggested that it might be interesting to see if it would be possible to determine how Black's treatment of the 'new chemistry' changed over the years. For various reasons the investigation was not actively pursued, but the search for manuscripts continued in the hope that perhaps the project would be revived. In the meantime Dr. Douglas McKie was publishing his series of papers in the *Annals of Science* on the manuscripts, and in one of the articles he indicated that he planned to discuss later how Black treated Lavoisier's work. Unfortunately, he was not able to complete this and only a comment to the effect that Black introduced Lavoisier's ideas in a non-systematic fashion occurs in a related paper.

The search for manuscripts began with a visit to the University of Edinburgh and the National Library of Scotland in 1955. Lack of time was (and has remained) a factor and so the manuscripts were listed and quickly examined for content, although microfilm copies of several were obtained. No other archives were contacted during that visit. Later, use was made of the listings of manuscript collections (see references 15, 16 and 17). A query was sent to *Isis* and the response brought to light a number of manuscripts in the United States as well as possible locations in Great Britain. These suggested locations were contacted by letter with a few positive replies. Of major assistance has been the information and encouragement furnished by Professor Henry Guerlac, Dr. William Smeaton and, most recently, Dr. R.G.W. Anderson. The latter's list of some fifty manuscripts led to the examination in 1979 of eight 'new' items. Book dealers' catalogues and the dealers themselves have been of assistance. For example Henry Sotheran and Company of London were most helpful in tracing manuscript 71 in this survey. The literature and periodicals have been of some use (especially for manuscripts 86 and 87).

Most of the manuscripts have been seen but not necessarily examined in detail. Those not seen include Numbers 1 (photocopy courtesy of Dr. R.G.W. Anderson, Science Museum, London), 2, 5 (photocopy courtesy of Dr. Ian H.C. Fraser, Archivist, University of Keele), 6 and 7 (information and photocopies of a few pages furnished by Mr. J.V. Golinski, University of Leeds), 16, 28, 54 (published by ICI, 1966), 68, 70, 82, 83 (photocopy of

selected sections courtesy of Miss Diana J. Dyason, University of Melbourne), 84, 85 (microfilm and a few photocopied leaves courtesy of M. Ron, The Jewish National & University Library, Jerusalem), 86 and 87.

The list has been arranged by location, with Great Britain divided into England, Scotland and Wales, followed by the United States, arranged alphabetically by state and city, and finally other countries in alphabetical order. Within the individual descriptions, if the manuscript has a title page it is transcribed for the first volume only. The designation 4° or 8° is rather generalized to refer to size and does not necessarily represent gatherings in the usual sense. Pagination or foliation is not always given in detail. In many cases all of the blank leaves and misnumberings are not listed. If only one side of the leaf is written on, the verso is frequently used for notes, even if this is not specifically stated. Generally, if the manuscript includes lectures on thermometers and temperature scales, a chart listing boiling points, freezing points, greatest cold, etc. is present, either as an inserted folding sheet or as a list in the text. Drawings of apparatus are included in some manuscripts but these are not specified in all cases. Several copies include the printed sheets beginning "Preparations of Mercury . . ." and "Preparations of Antimony . . .": these are Numbers 3, 9, 13, 22, 24, 27, 30, 37 (Mercury only?), 40, 46 (the only one in Latin, all of the others are in English), 64, 80 and 81. Dating has been obtained primarily from dated lectures and from the date in the title. The latter method is obviously not always trustworthy and so in some cases, where time permitted, internal evidence was sought, such as the reference in lecture 3 to James Price converting mercury to gold (1782), and in lecture 84 to C.L. Berthollet's experiment in June of 1787 producing "dephlogisticated muriatic acid". Approximately 120 references which might be of use in dating are known at present. One point, however, that is of no use in dating the manuscripts occurs at times in lecture 17 (evaporation under reduced pressure) where we find reference to "Robison who is Lecturer in Chemistry at Glasgow . . .". Robison left Glasgow in 1774 and settled in Edinburgh in September of that year. This would seem to date a manuscript as being before 1774. However, this statement occurs in manuscripts as late as 1788. Also, it appears that much of the material on heat was not changed significantly from year to year and that the 'master copies' which were available to the students for copying were not corrected in the case of 'minor points'. Dating of many of the manuscripts needs more detailed checking; thus for example, manuscript 14 has the date Dec. 15, 1775 for

Figure 6. Joseph Black lecturing in Edinburgh in session 1767-68: ink sketch by a student, Thomas Cochrane, in manuscript 54.
(Reproduced by permission of the Andersonian Library, University of Strathclyde)

lecture 29, but there are references in the text to discoveries made in 1780 or later.

The list is presented as a 'first effort' and is not intended to be a complete description of the 87 known manuscripts. We hope it will stimulate the 'discovery' of other manuscripts. My thanks are due to all of the librarians and archivists in the various institutions for their unstinting and kind assistance. I owe thanks to the holders of manuscripts in private collections for either making the manuscripts available for examination or for sharing information relating to them. The errors in the list, which I hope are few, are strictly my own.

REFERENCES CITED IN THE SURVEY

1. R.G.W. Anderson *The Playfair Collection [at the Royal Scottish Museum] and the Teaching of Chemistry at the University of Edinburgh 1713-1858* (Edinburgh 1978)
   a) p 103, b) pp 35-42, c) p 36 note 17.

2. M.P. Crosland 'The Use of Diagrams as Chemical 'Equations' in the Lecture Notes of William Cullen and Joseph Black' *Annals of Science. 15* 75-90 (1959),
   a) pp 85-88, b) pp 84-85, c) pp 80, 83, d) p 82, e) p 85.

3. P.M. Hamer *A Guide to Archives and Manuscripts in the United States* (New Haven, Conn. 1961)
   a) p 538, b) p 636.

4. *Index-Catalogue of the Library of the Surgeon-General's Office United States Army, Authors and Subjects* Vol II (Washington, D.C. 1881) p 78: Lectures on chemistry MS by James Bense (sic) 2v 8° (Edinb. 1795-6?).

5. A. Kent (editor) *An Eighteenth Century Lectureship in Chemistry* (Glasgow 1950)
   a) pp 88-89, b) pp 89-91, c) p 97 note 18.

6. Douglas McKie 'On Thos. Cochrane's MS. Notes of Black's Chemical Lectures, 1767-8' *Annals of Science. 1* 101-110 (1936).

7. Douglas McKie 'On Some MS. Copies of Black's Chemical Lectures – II' *Annals of Science 15* 65-73 (1959) p 73.

8. *Ibid*, III, *Annals of Science 16* 1-9 (1960) pp 7-9.

9. *Ibid*, IV, *Annals of Science 18* 87-97 (1962) p 97.

10. *Ibid*, V, *Annals of Science 21* 209-255 (1965)
    a) pp 252 –, b) pp 209-255.

11. *Ibid*, VI, *Annals of Science 23* 1-33 (1967)
    a) p 2– , b) pp 32-3, c) p 33, d) pp 1-33.

12. Douglas McKie & N.H. de V. Heathcote 'William Cleghorn's *De Igne* (1779)' *Annals of Science 14* 1-82 (1958) p 67.

13. Douglas McKie & D. Kennedy 'On Some Letters of Joseph Black and Others' *Annals of Science 16* 129-170 (1960) p 166.

14. National Register of Archives Personal Index (London): a computer printout in the Manuscript Department of the British Library.

15. *The National Union Catalog of Manuscript Collections 1959-61* (Ann Arbor, Mich. 1962): MS 60-2820, Hist. Soc. Pennsylvania, Medical Notes 1785-86, 6 volumes, William Martin (1765-98).

16. *National Union Catalog of Manuscript Collections. Catalog 1977* (Washington, D.C. 1978): MS 77-1414, Cornell University Hist. Sci. Coll., in the Lavoisier Collection (no details given).

17. *National Union Catalog Pre-1956 Imprints* Vol LIX (London 1969).

18. D.R. Oldroyd 'Two little known Copies of Black's Lecture Notes' *Annals of Science 29* 35-37 (1972).

19. J.R. Partington *A History of Chemistry* Vol III (London 1962) p 131 note 4.

20. J. Read *Humour and Humanism in Chemistry* (London 1947)
    a) pp 162-163, facsim., b) pp 164-166, facsim.

21. M. Ron *From Alchemy to Atoms An Exhibition of Books, Documents, Mss., etc. on the History of Chemistry and Chemical Technology from the Sidney M. Edelstein Collection* (Jerusalem 1978) p 31, No 49.

22. R.E. Schofield *Mechanism and Materialism* (Princeton, New Jersey 1970) p 186 note 25.

23. L.C. Witten II and R. Pachella *Alchemy and the Occult A Catalogue of Books and Manuscripts from the Collection of Paul and Mary Mellon given to Yale University Library* Vol IV (New Haven, Conn. 1977)
    a) pp 658-661, facsim., b) pp 678-681, facsim., c) pp 705-707 facsim.

24. H. Zeitlinger *Bibliotheca Chemico-Mathematica* (London 1921) p 318, item 6249.

25. H. Zeitlinger *Sotheran's Bibliotheca Chemico-Mathematica . . . First Supplement* (London 1932)
    a) p 358, item 5061, b) p 358, item 5062, c) p 358, item 5063.

26. H. Zeitlinger *Sotheran's Bibliotheca Chemico-Mathematica . . . Second Supplement* (London 1937)
    a) p 656, item 10269, b) p 656-7, item 10270, c) p 657, item 10272.

Manuscripts

GREAT BRITAIN, ENGLAND

1. **Birmingham: Birmingham Assay Office**
   MS 98/103 1775-76

   "Theory of Heat; abstracted [by Charles Darwin] from Notes taken during the Sessions 1775-6 at Glasgow (sic) University"

   Part of a 4° volume of miscellaneous notes; both sides, one hand; pp 1-9.

   Lectures: Approximates to numbers 13-18. They include Capacity for Heat, Fluidity, Latent Heat and Evaporation.

   Charles Darwin, son of Erasmus Darwin (1731-1802), transcribed the notes for Matthew Boulton.

   Ref: 22

2. **Cambridge: Cambridge University Library**
   Add. MS 7736 n.d. (1781-82?)
   (Wollaston Papers)

   No title

   Four volumes, approximately 8°, Vols I-III are 18.4 x 11.4 cm, Vol IV is 20.3 x 16.5 cm; Vols I-III are both sides except for an occasional mistake when Tennant turned over two pages, Vol IV is one side. All volumes are in Tennant's own handwriting. Vol I: 58 pp; II: 58 pp; III: 28 pp + 15 blank leaves; IV: 14 pp (p 15 has 4 pen and ink sketches of Papin's digester and 3 kinds of retort). The balance of the total 94 pp is blank.

   Lectures: 34-124 in Vols I-III. Vol III starts with 3 pp of symbols and tables of elective attractions with notes. Vol IV contains "Lectures on Heat".

   Smithson Tennant (1761-1815) entered Edinburgh University in 1781, and subsequently graduated from Cambridge where he was later professor of chemistry (1813-15). He was elected FRS in 1785, and from 1800 was a partner of W.H. Wollaston.

3. **Cambridge: Private collection of Professor Duncan McKie, Jesus College**
   1770-71

   "A Course of Lectures on Chymistry by Joseph Black M.D. Professor of Chymistry in the University of Edinburgh Volume First".

   Four volumes 4° 23.8 x 18.4 cm; one side, one hand; I: contents ff 1-11, text ff 2-290 (185-6 are blank); II: contents ff 1-7, text ff 1-287; III: contents ff 1-4, text ff 1-237; IV: contents ff 1-6, text ff 1-278, index ff 1-13.

   Lectures: 1-114 (as numbered); introductory lecture not present. Verso f 277 in Vol IV: "Concluded on Saturday April 27th 1771".

   Alexander Anderson of Bourtie (enrolled 1770-71): his class ticket dated 31st Oct. 1770 in Vol I; also a class ticket for Natural Philosophy, Edinb. Oct. 1767, Jas. Russell the instructor. Bookplate: John Leith Ross of Arnage and Bourtie.

   Refs: 9, 11a, 12

4. **Cambridge: Private collection of Professor Duncan McKie, Jesus College**
   1790-91

   No title

   One volume 4° 22.5 x 18.4 cm; both sides in two or more hands; pp 1-590, 2 pp blank, 2 pp index (incomplete).

   Lectures: 1-115. Some are dated, e.g. lecture 38: Friday Jan. 17th 1791.

   Wm. Allen (1770-1843).

   Refs: 9, 10a, 13

5. **Keele: University of Keele, The Library**
   Wedgwood MS 28409-39 1766-67
   (property of Messrs Josiah Wedgwood & Sons Ltd., Barlaston, Stoke-on-Trent)

   "Lectures on Chemistry by Joseph Black M.D. Prof. Chem. in University of Edinburgh 1766-7" (caption title p 143)

   Part of a 4° volume containing other notes; both sides, one hand; pp 143-154 (on pp 152-3 "Contents of the above first volume", p 154 extract of a letter from Mr. More, April 15 1775, on Latent Heat).

   Lectures: Not numbered, but a part of the lectures on heat.

   Alexander Chisholm. Chisholm was the assistant of William Lewis (1708-1781) and later of Josiah Wedgwood (1730-1795).

   Ref: 22

6. **Leeds: University of Leeds Library**
   MS 34 (Special Collections) 1786-87

   "Lectures on Chemistry By Joseph Black. M:D: Prof. of Chem: in the University of Edinburgh Vol: 1 [quotation from Virgil] 1786-7."

   Five volumes, numbered 1, 2, 3, 9 and 10, each with a title page, 8° 14.5 x 12.0 cm; one side, one hand; I: 10 blank leaves, ff 1-185, 9 blank leaves; II: 9 blank leaves, ff 1-174, 7 blank leaves; III: 4 blank leaves, ff 1-203, 8 blank leaves; IX: 16 blank leaves, 180 ff; X: 210 ff. Only Vols I-III have tables of contents.

   Lectures: 1-40, 102-110, "105-107" (numbers repeated) and 4 unnumbered lectures.

   No name
   Part of the collection formed by Alfred Chaston Chapman.

   Refs: 25b, 26c

7. **Leeds: University of Leeds, Medical Library Historial Collection**
   n.d. (1772-73?)

   No title

   One volume 4° 23.5 x 17.0 cm; both sides, one hand; 6 blank pp, pp 1-709, 3 pp (Contents), 3 blank pages.

Lectures: 1-62.

James Tatham's (sic), on the third of the un-numbered preliminary pages. The spine of the volume bears the title: "Latham (sic) Chemistry Lectures". The volume has only recently been identified as being Joseph Black's lectures. James Tatham apparently did not graduate M.D. from Edinburgh.

Presented to the Leeds School of Medicine by Edward Thompson in 1880.

8. **Liverpool: The University of Liverpool,**
   Sydney Jones Library
   MS 4/7   n.d.

"Minutes from a course of lectures on Chemistry. Read before the Students at Edingburg (sic) By Joseph Black M.D. Prof: Chem: Vol III"

One volume 4° 25.2 x 19.1 cm; one side, two hands; leaves are numbered using a consecutive series of odd numbers: ff 11-759 (the volume has been systematically damaged by the removal of alternate pairs of leaves), following f 759 are several blank leaves and stubs then an 'Index to Contents Volume 1st' on one leaf (both sides), 2 stubs, 2 leaves 'Contents V. 2' (both sides), 2 stubs, 2 leaves ruled but blank, 2 stubs, 2 leaves ruled but blank, 2 stubs. There are a few errors in the numbering of the leaves.

Lectures: "85-118" (approximately half of the text is present). On the verso of f 195 is a note relating to pewter signed R. 1784.

Gift of Hugh Rathbone.

Ref: 14 (entry 12597(67), dated as c1775)

9. **London: The British Library,**
   Department of Manuscripts
   Add. MS 52495   1769-70

No title

One volume 4° 20.4 x 16.3 cm; both sides, one hand; 408 pp + 4 leaves torn out.

Lectures: Approximately 130 present, not numbered; some have day and month without the year; on Dec. 8 (p 69) there is a note to the effect that the lectures ceased until Jan. 2 1770 because Dr. Black was injured in a fall from his horse.

No name

Refs: 11a, 14 (mistakenly gives the call number as Add. MS 52496)

10. **London: The British Library,**
    Department of Manuscripts
    Add. MS 59843-45   n.d.

No title

Three volumes sm 4° 19.5 x 15.5 cm; one side, one hand; leaves are not numbered; I: 384 ff; II: 349 ff; III: 449 ff (the count includes a few blank leaves).

Lectures: 1-11 are numbered, the remainder are not; total 118?

No name

This MS was formerly at the Patent Office, London.

Refs: 2a, 7

11. **London: The Chemical Society**
    MS E 151i   1773-74

No title

Two volumes 4° 23.8 x 18.7 cm; both sides, two (?) hands; V: pp 1-275; VI: pp 1-303.

Lectures: 80-118. Lecture 118 dated May 2, 1774.

Bookplate of H.E. Roscoe in Vol V, that of John Glover in Vol VI.

Ref: 19

12. **London: The Chemical Society**
    MS E 151i   n.d. (1773-74?)
    (shelved with No. 11)

"Lectures on chemistry By Joseph Black M.D. & P. Edinburgh Vol. I."

Four volumes 4° 24.7 x 19.4 cm; both sides, one hand; I: 379 pp; II: 436 pp; III: 388 pp: IV: 360 pp (numbers are approximate).

Lectures: 1-99. Where they overlap with those of No. 11 they are identical, word for word.

Bookplate of H.E. Roscoe in Vols I & III, that of John Glover in Vol II.

Ref: 19

13. **London: The Pharmaceutical Society of**
    Great Britain
    MS 091 BLA   1773-74

"Lectures on Chemistry by Joseph Black. M.D. Professor of Chemistry in the University of Edinburgh Vol. I A.D. 1773-74"

Four volumes 4° 23.8 x 18.7 cm; both sides, one hand; I: 4 leaves (1st with a name, 2nd blank, 3rd title, 4th with a name), pp 1-313, tipped in is a leaf with a note; II: 4 leaves (as in Vol I except 2nd is title and 3rd and 4th are blank), pp 1-94, '94-95', 2 blank leaves, pp 95-309, 9 blank leaves, 3 pp Index; III: 4 leaves (as in Vol I except 4th is blank), pp 1-308, 13 blank leaves, 1 leaf (both sides) Index; IV: 4 leaves (as in Vol III), pp 1-62, '62-63', 63-84, '84-85', 85-230, 34 blank leaves, 1 leaf (both sides) Index.

Lectures: 1-118. Several are dated, e.g. lecture 70: Febry 17th 1774.

I. Cross on title pages, and also I. Crosse on Vol I leaf 4. Edme Graves on first leaves, but Edme Greaves in Vol III.

14. **London: The Royal Society**
    MS 147-149   1775-76

"A Course of Lectures on the Theory and Practice of Chemistry by J. Black M.D. Professor of Chemistry in the University of Edinburgh"

Three volumes 4° 22.3 x 17.5 cm; Vol I for the most part is on both sides, Vols II & III are one side, several hands; I: pp 1-362, ff 363-374, pp 375-382, ff 383-426; II: ff 1-394; III: ff 1-421.

Lectures: 1-118. Lecture 29 dated Dec. 15, 1775. Vol II lecture 69 states a new edition of Crawford is to be published soon (1788 was the date of the second edition); Vol III lecture 97 refers to work of Lavoisier (c1784).

No name

Ref: 11c

15. **London: The Royal Society**
    MS 144-146   1778-79

No title

Three volumes 4° 22.9 x 18.2 cm; both sides, one hand; I: 1 leaf Note*, 198 ff; II: 215 ff; III: 224 ff.

* "Lectures read by me at the University of Edinburgh in 1782. and presented by me to Mr Sheriff Gordon as a mark of my esteem for him as a friend. J. Black MD LL.D Edin." This note may not be in Black's hand.

Lectures: 1-118. Lecture 29 dated Dec. 15 1778.
Bookplate of Alexander Gordon.
Presented to the Royal Society 22 May 1947 in accordance with the wishes of the late Dr. Alexander Scott.

Ref: 11b

16. **London: Science Museum Library**
    MS 420   n.d. (c1773)

"Chemistry" (caption title p1)

One volume 4° 23.1 x 18.7 cm; both sides, one hand; pp 1-469.

Lectures: 1-44, 45 (incomplete). The handwriting in this manuscript and No. 17 appear to be identical, making this Vol I and No. 17 Vol II of a set wanting Vol III.

No name
Bookplate inside front cover with family name "Clifton" and the motto "Tenez le Droit".

17. **London: Private collection of the late**
    Professor Franz Sondheimer,
    University College, London
                                        n.d. (c1773)

No title

One volume 4° 23.2 x 18.2 cm; both sides, one hand; pp 470-800, 800-821, 822 (in pencil), 823-911 not numbered, 912 (in pencil) is blank.

Lectures: Last part of 45, and lectures 46-86. See No. 16.

No name

18. **London: University College London,**
    D.M.S. Watson Library
    MS Add 96   n.d. (c1773)

"Lectures on Chimestry (sic) by Joseph Black M.D." (caption title f 1)

Six volumes 4° 23.2 x 18.4 cm; one side, three hands; I: 195 ff; II: 226 ff; III: 232 ff; IV: 236 ff; V: 206 ff; VI: 203 ff.

Lectures: 1-118. From the contents a likely date is c1773.

No name

Refs: 7, 8, 10b, 14, 19

19. **London: Wellcome Institute for the**
    History of Medicine Library
    MS 1219-1227   1766-67

"Notes of Dr. Black's Lectures. The Introduction & History are omitted" (caption title f 1)

Nine volumes 8° 16 x 11 cm (not uniform), one side, except for the first four leaves of Vol I, one hand; foliated continuously, 687 leaves total.

Lectures: 126. Dated 17 Nov. 1766 – 19 May 1767.

Sir Charles Blagden (1748-1820), enrolled 1766-67, 1767-68; M.D. 1768.

Ref: 11d

20. **London: Wellcome Institute for the**
    History of Medicine Library
    MS 1228   1776-77

"Lectures on Chemistry Begun 30th October 1776 By Dr. Joseph Black Edinburgh"

One volume 4° 20 x 18.5 cm; both sides, one hand; 494 pp.

Lectures: 1-26, part of 27.

No name. A note in pencil "His original MSS", but the lectures are not in Black's hand.

21. **London: Wellcome Institute for the**
    History of Medicine Library
    MS 1229-1232   1777-78

"Dr. Black's Lectures of X^y begun 3 Novr 1777 Edin. Robert Perceval"

Four volumes 8° 18.5 x 11.0 cm; generally one side, one hand; a total of about 290 leaves used.

Lectures: 118? They are not numbered.

Robert Perceval (1756-1839), enrolled from 1777 to 1780; M.D. 1780. He was the first professor of chemistry at Trinity College Dublin.

Ref: 1a

22. **Manchester: John Rylands University Library**
    of Manchester
    MS CH.B 106   1768-89

"Chemistry Dr. Black 1768 Novr. 2d The course began, prelim^y lect^r finished on Tuesday Novr 8"

Eight volumes 8° 18.4 x 11.4 cm; Vols I-V one side, Vols VI-VIII both sides, one hand; approximately 1376 written pages total.

Lectures: 137, with the first lecture omitted.

Robert Dobson (signature in Vols I & VII) attended 1768-69, 1769-70, and graduated M.D. and M.A. 1771.
In Vol VII below the signature of Dobson is another: Geo. Freckleton Trin. Col. Cam.

Ref: 11d

23. **Manchester: John Rylands University Library**
    of Manchester
    MS CH.B 107   n.d. (1769-70?)

No title

One volume 8° 18.5 x 11.5 cm; both sides, one hand; 118 pp + 22 blank leaves.

Lectures: Not numbered. Covering Theory of Quicklime to Brass in the lecture on Copper.

No name, but the hand is that of Robert Dobson (see No. 22).

24. **Manchester: John Rylands University Library of Manchester**
MS CH.B 108   1774-75

"A Course of Lectures On Chymistry. Delivered by J. Black M.D. Professor of Chymistry In the University of Edinburgh Anno 1774-75 Vol Ist"

Four volumes 8⁰ 18.6 x 11.2 cm; one side, one (?) hand; I: title/blank, Contents/blank, ff 1-319; II: ff 320-629; III: ff 630-938; IV: ff [939-1010], ff 748-980; some blank leaves and some errors in numbering.

Lectures: Not numbered. Pharmacy is omitted.

No name
Owned once by Geo. Wilson M.D., bought by H.B. Hodgson and then given to H.E. Roscoe.

25. **Manchester: John Rylands University Library of Manchester**
MS CH.B 109   1795-96

No title

Two volumes 8⁰ 18.7 x 11.0 cm; both sides, one hand; I: laid in folded sheet iv pp Dr. Black's definition of chemistry, pp 1-473, 24 pp Syllabus of Dr. B's lectures on chemistry. Edinb. 1795-96; II: pp 475-713, 8 blank leaves then about 150 leaves of medical case histories.

Lectures: Starts with Heat; perhaps 130 lectures in all, but not numbered. On p 713: "Conclusion of the Course April 30th 1796 Left Edinb. beginning of December 1796." This was a course given jointly by Black and Thomas C. Hope. The lecturer is indicated for many of the sessions. See also No. 46.

No name

26. **Manchester: John Rylands University Library of Manchester**
MS CH.B 110   n.d.

"Lectures on Chemistry by Joseph Black M.D. Vol. VI."

One volume sm 4⁰ 20.3 x 12.3 cm; one side, one hand; title, ff i-x Index, ff 1-261.

Lectures: Last part of 89, and lectures 90-105.

Thos. Henry: signature on free end paper. Thomas Henry (?1734-1816). Presented 4 Mar. 1851 by Dr. Charles Henry.

27. **Manchester: The University of Manchester Institute of Science and Technology**
MS J660 B2   1769-70
(Joule Collection)

No title

Three volumes sm 4⁰ 19.3 x 15.0 cm; both sides, one hand; I: 384 pp, 1 blank leaf, 4 pp Table of Contents; II: 380 pp, 4 pp Table of Contents (last page blank); III: 318 pp (no Table of Contents).

Lectures: ? − 137. Only in Vol III are the lectures numbered. Starts with the History of Chemistry.

Henry Richardson Junr. Vol III is signed and dated Edinburgh 1769. Richardson was enrolled 1769-70 and 1770-71.

Ref: 11d

28. **Newcastle: Newcastle University Library**
Med Coll. 540.4 (Acc V5592-3)   n.d.

No title

Two volumes 4⁰ 27.4 x 22.0 cm; both sides, several hands; I: 306pp, 12 blank leaves; II: 391 pp, 40 blank leaves.

Lectures: 1-137.

No name
Stamp of Newcastle Infirmary Medical Library and 'Presented by Mr. Emerson Charnley 1819' on page 1 of each volume. Previously offered for sale (item 39) as 'Black's Chemical Lectures, in MS. 2 vols. half bound, 12s.' in Charnley's *A Catalogue of books, including the libraries of the late Mr. Horn and Mr. Burnett, surgeons, Newcastle . . . 1817.*

GREAT BRITAIN, SCOTLAND

29. **Aberdeen: Aberdeen University Library, Kings College**
MS 472 (Skene Collection)   n.d.

"Notes on Chemistry by Dr. David Skene"

One volume 8⁰ 18.6 x 14.9 cm; one side, one hand; laid in at front a folded sheet 5 pp of text + 3 pp blank, 1 leaf title, ff 1-350 (f 95 is also numbered 110, this double numbering occurs in several places; ff 18-20 are blank).

Lectures: Only 1-11 are numbered.

David Skene (1731-1770).
Bookplate of Alexander Thomson of Banchory.

30. **Edinburgh: National Library of Scotland**
MS 3533-34   1767-68

"Elements of the Theory and Practice of Chymistry. delivered by Joseph Black Profess^r of Chymistry at Edinb^r In two volumes Vol 1st 1767"

Two volumes 4⁰ 23 x 18 cm; both sides, one hand with many notes and corrections; I: pp i-xii, 1-262; II: pp i-xii, 1-380.

Lectures: Not numbered or dated.

N. Dimsdale: signature on title of each volume. Nathaniel Dimsdale attended 1767-1771 and graduated M.D. 1771.

Ref: 11d

31. **Edinburgh: National Library of Scotland**
MS 5725   1769-70

"Chemistry by Dr. B. begun Octr 1769"

One volume 8⁰ 15.5 x 9.2 cm; from lecture 30 it is for the most part both sides, one hand; ff 15-126 (there are some old numbers in ink and some mixup of the leaves).

Lectures: 7-53.

William Stewart. The notebook was first used by his father, James Steuart, Keeper of the King's Wardrobe, then inverted and used by William for notes on arithmetic, Greek, shorthand and chemistry.

32. **Edinburgh: National Library of Scotland**
    MS 8487   n.d.

No title

One volume 8° 18.5 x 11.0 cm; one side, two hands; ff 1-262 (less 133).

Lectures: 6-20.

No name

33. **Edinburgh: Royal College of Physicians**
    MS Black 4   1778

"Lectures on Chemistry" (on free end paper)

One volume 8° 22.5 x 12.5 cm; one side in short-hand; ff 1-51 (ff 52 & 53 are blank, then begins "Practice of Medicine", there are a few blanks in the text). These notes are in an inverted set of William Cullen Clinical Lectures.

Lectures: Not numbered. Starts with Crystalline Earths, Jan. 1778, and ends with Metals. There is no real order, e.g. Copper appears on ff 4-6.

William Keir (?): ascription in College Library catalogue.

34. **Edinburgh: Royal College of Physicians**
    MS Black 3 (formerly M 9.42-43)   n.d.

No title

Two volumes 4° 23.0 x 18.6 cm; both sides, one hand; I: 632 pp; II: 670 pp, 2 blank leaves.

Lectures: 1-118.

No name

Ref: 2b, where Dr. Crosland gives a date of 1786 (as does the College Library catalogue). The text, except for a few additions (e.g. Vol I p 252 "in ye year 1781 ye cold was observed at Glasgow to be 14° below O.") appears to be of an earlier date.

35. **Edinburgh: Royal College of Physicians**
    MS Black 2   1785-86
    (formerly M 9.44-49)

No title

Six volumes 8° 20.2 x 15.8 cm; one side, seems to be one hand; I: 237 ff, 2 blank leaves; II: 223 ff; III: 285 ff; IV: 270 ff; V: 214 ff; VI: 295 ff. Leaves have been removed from some volumes without affecting the text.

Lectures: 1-118. Lecture 1 dated 15 Nov. 1785, lecture 95 May 1st 1786.

James Scott: signature on fly leaf Vol VI. Scott enrolled 1784-5, 1785-6; there is another (?) James Scott 1785 to 1789.

Refs: 2b, 14 (entry 16015, dated as c1770? – c1785: this may however refer to No. 34)

36. **Edinburgh: Royal College of Physicians**
    MS Black 1   n.d. (c1788)
    (formerly M7.33-40)

"Lectures on Chemistry by Joseph Black M.D."

Eight volumes 8° 21.2 x 12.6 cm; one side, two or more hands; I: 316 ff; II: 285 ff; III: 246 ff; IV: 277 ff; V: 271 ff; VI: 333 ff; VII: 242 ff; VIII: 210 ff. Each volume starts with a title and 3 blank leaves, they have varying numbers of blanks at the end and a few in the text. Gatherings of 8 leaves are 'signed' with numerals.

Lectures: 1-118. This MS seems to be c1788. Lecture 15 contains a reference to the "late Dr. Irvine of Glasgow . . ." (William Irvine died July 9, 1787) and lecture 87 has a reference to H.B. de Saussure's *Voyages dans les Alpes* (published in 1786).

A. Duncan: signature in Vol V on the fly leaf. Part of the bequest of Dr. Andrew Duncan *Senior* (1744-1828).

Ref: 2c

37. **Edinburgh: University of Edinburgh Library**
    MS Dc.10.1   1770-71

"Lectures of Chemistry By Joseph Black Professor of Chemistry in the University of Edinburgh" (This is followed by a second title: "Chymical Lectures By Dr. Joseph Black Professor of Chymistry in the University of Edinburgh A.D. 1770 James Johnson Edinburgh Dec 10th 1774 finished Jan 8th 1775 per me J.J. Jacobus Johnson")

One volume 4° 24.3 x 20.1 cm; both sides, one hand; pp i-iii, 1-34, 8 pp, pp 35-505, p 506 note stating that the true first lecture was missed, p 507 index (only two entries), pp 508-510 are blank.

Lectures 1-120 (as numbered).

James Johnson, Surgeon, Lancaster: card as book-plate. On a paper tipped in: "Copies of lectures taken on the spot in shorthand by James Johnson, formerly of Lancaster, presented to the Library of the University of Edinburgh by his grandson Christopher Johnson of Lancaster May 4th 1876".

Refs: 2d, 17 (entry NB 0520230: microfilm copy at Cornell University, Ithaca, N.Y.)

38. **Edinburgh: University of Edinburgh Library**
    MS Dc.8.174   1771-72

"Lectures on Chymistry by Joseph Black M.D. Edinr Novr 1771"

One volume 8° 18.0 x 10.7 cm; one side, one hand; ff 1-6 (blank except for notes regarding provenance and f 6 which is the title), 7-89 (ff 7-89 also numbered as pages 333-497). The volume is numbered '9'.

Lectures: 50, 52-78.

John Hill.
Bookplate of James Nairne of Claremont. Notes on ff 1-5 include: "M.S.S. of my learned Father-in-law saved by me from being Cancelled – Circa 1808 or 9 JsN" and "M.S.S. Dr. John Hill Professor of Humanity in the University of Edinburgh preserved by his son in law James Nairne Esq. Presented to John Cook his Grand Nephew 4 November 1890".

39. **Edinburgh: University of Edinburgh Library**
    MS Dk.342-45   1774-75

"Lectures on Chemistry Delivered by Joseph Black M.D. Professor of Chem in the University of Edinburgh 1774 Vol 1st"

Four volumes 4° 23.1 x 17.5 cm; both sides, two or more hands; I: title, pp 1-460, 4 leaves index; II: title, pp 461-911, 4 leaves index; III: title, pp 912-1368, 5 leaves index; IV: title, pp 1369-1812, 5 leaves index; laid in Vol I large drawings of (1) Athanor furnace and (2) Reverberatory furnace; in Vol II a large drawing of Black's furnace.

Lectures: 1-118.

At the end of the index of Vol I are the initials A.G.; vol III end paper: "Ex Officina Jacobi Bon Ayr 1782 James Bon Surgeon of Ayr" (a James Bon was enrolled 1775-6 and 1776-7).
Bookplate: Carnegie Library, Ayr.

Ref: 14 (entry LA 1957)

40. **Edinburgh: University of Edinburgh Library**
        MS Dc.3.11-13   1775

No title

Three volumes 8° 19.0 x 14.8 cm; both sides, one hand; I: pp 1-426, 1-26; II: pp 1-446, 1-8; III: pp 1-89, 1-211 Appendix, 1-66, 2 pp Supplement, 14 blank leaves, pp 1-54 Index. There are occasional intercalated leaves used for notes, etc; these are not numbered.

Lectures: 1-57. An abbreviated course. Starts with a paragraph defining chemistry then takes up Heat and ends with Animal Substances. Lecture 1 is dated 13 June 1775, lecture 54 is dated 19 Dec. 1775, and lecture 57 is 22 Dec, without the year. Vol I p 123: "Note What follows on the Chymical Apparatus is taken from the notes of a Gentleman who attended Dr. Black's Ordinary Course."

Alexander Law: signature Vol III p 1.

Ref: 17 (entry NB 0520259: microfilm copy at Cornell University, Ithaca, N.Y.)

41. **Edinburgh: University of Edinburgh Library**
        MS Dc.2.84   1781-82

"Lectures on Chymistry By Dr. Black 1781"

One volume (Vol I) 4° 23.0 x 18.2 cm; one side, one hand; title, ff 1-295, pp 1-8 index, at the end is a leaf numbered in pencil 11 and 12 (11 is blank, 12 has a note relating to p 46).

Lectures: 1-75. Lecture 75 Bitumens ends with jet and coal.

No name
Presented by Dr. W.K. Dickson, 1942.

42. **Edinburgh: University of Edinburgh Library**
        MS Dc.10.9^1-9^7 1782-83

"Dr. Black's Lectures Taken Down Anno Domini 1782:3 Begun 30 Octr 1782 Ended 3 May 83 pretty Exact Corrected AD 83-84 Lectures Vol. 1st 1782" (this is inverted on the verso of f 103 in Vol I; there is another 'title' at the start which repeats some of the above.)

Seven volumes 8° 18.5 x 11.5 cm; one side, two hands; I: ff 1-103; II: ff 1-92; III: ff 1-92; IV: ff 1-106; V: ff 1-61; VI: ff 1-95; VII: ff 1-65 + 29 blank leaves.

Lectures: 1-57, 62-64(?), 74-131 (as numbered). Some are dated, e.g. Vol V f 58 lecture 98, April 1st 1783. There are indications of evening lectures.

Thos Hope: signature on the title page of Vol IV (also very faintly James Noll Edinb). Thomas Charles Hope (1766-1844) attended courses 1782 to 1787. Part of the Hope bequest (see also No. 45).

Ref: 1b

43. **Edinburgh: University of Edinburgh Library**
        MS Dc.2.41   1782-83

No title

One volume 4° 23.7 x 17.6 cm; both sides, one hand; pp 1-149, [150-376].

Lectures: 1-59. Lecture 1 dated Oct 30 1782. The folding temperature scale has the addition: "now found mercury congeals so low as 45 below 0 1784".

No name
Presented by the Ulster Medical Society 20 Oct. 1914.

44. **Edinburgh: University of Edinburgh Library**
        MS Gen 2031   1783-84

"Notes of Dr. Joseph Black's Lectures on Chemistry by John Caw Vol. IInd 1783".

One volume 8° 18.2 x 11.5 cm; both sides, one hand; title, pp 183-368 (pp 317-333 and 363-367 are blank).

Lectures: 27-60, less 53 and 54. Lecture 27 dated 22 December 1783.

John Caw, enrolled 1783-84: a second name John Young Caw also appears. Presented to the University of Edinburgh Jan. 23, 1978 by Miss C.R. Cheesman.

45. **Edinburgh: University of Edinburgh Library**
        MS Dc.10.9^8 1787?

No title

One volume 8° 18.6 x 11.4 cm (bound 'uniformly' with No. 42); one side, several hands; 5 leaves (one with a name, one with a note regarding the definition of chemistry, the rest are blank), ff 2-54, 10 blank leaves (one has a note concerning salts).

Lectures: 1-118. A course outline.

Jas M Cartney College Edr 1787.
Owned by T.C. Hope, and part of the Hope bequest; see No. 42.

46. **Edinburgh: University of Edinburgh Library**
        MS Gen 48D   1796-97

"Chemistry by Drs. Black and Hope" (caption title p 1)

One volume 8° 18.0 x 14.7 cm; both sides, one hand; 279 ff (f 279 verso is blank).

Lectures: 2-?; may be over 135 total. Starts with lecture 2 dated Oct. 31 1796. Many of the lectures indicate the name of the instructor.

No name
From Thomas Toon, Westminster, presented by Prof. A. Crum Brown to Chem. Libr. transferred 1960.

Ref: 1c

47. **Edinburgh: University of Edinburgh Library**
        MS Dc.8.48   n.d. (c1786)

No title

One volume (Vol II) 8° 19.7 x 11.8 cm; both sides, one hand; pp 308-620 (leaves seem to be mixed up but the pagination is continuous).

Lectures: Not numbered. Starts mid-sentence regarding Spirit of Wine ends with Milk (Ambergris is with the Bitumens; after 1787 Black discussed it under Balsams and Resins. On p 620 a reference to the New Dispensatory perhaps means the Edinburgh 1786 edition.)

William Robertson Surgeon Kelso 1797. Records show several Wm. Robertsons: one 1784-85, another 1795-97, etc.

48. **Edinburgh: University of Edinburgh Library**
    MS Dc.8.176   n.d.

No title

One volume (Vol VI) 8⁰ 20.0 x 12.8 cm; one side, one hand; ff 1-296, 7 leaves (mostly blank, verso of first has a few entries for an index, the verso of the second has 'chemical marks').

Lectures: 103-118. A note on the verso of f 92 cites Lavoisier and his opinion that water is composed of Empyreal and Inflammable airs (c1783).

C. Cotton (or Colton?): rubber stamp on f 1.

49. **Edinburgh: University of Edinburgh Library**
    MS Dc.2.42-43   n.d. (c1788)

No title

Two volumes (Vols II and III) 4⁰ 24.1 x 17.8 cm; both sides, one hand; II: pp 1-421; III: pp 1-425.

Lectures: 30-118. Lecture 86 notes "Berthollet performed the following experiment in June 1787 in the Acad. of Sciences of Paris . . .".

No name

50. **Glasgow: University of Glasgow, The Library**
    MS Gen 49/1-3   1773-74

No title

Three volumes 4⁰ 24.4 x 19.1 cm; both sides, one hand; I: pp 1-376, [377-492]; II: pp 1-456 (164 is blank); III: pp 1-269, 74 blank leaves.

Lectures: 1-118. At the end: "May 2, 1774 Finis J. Bruce Script". Some lectures are dated, e.g. lecture 102 at the end (Vol III p 52): "We procede upon Monday to the consideration of Gold. Saturday April 9th 1774."

James Bruce: see No. 69.
In each volume is the signature of Dugald Stewart (1753-1818) who was professor of moral philosophy at Edinburgh, 1785 to 1820. Also present is an inscription: "A. Bannatyne a present from his uncle (1816)" and a bookplate: Milheugh (i.e. John Millar of Milheugh (1735-1801) professor of civil law, Glasgow University, 1761-1801).

51. **Glasgow: University of Glasgow, The Library**
    MS Ferguson 173-174   1776-77?

No title

Two volumes 4⁰ 24.3 x 19.6 cm; both sides, one hand; I: pp 1-564; II: pp 1-570.

Lectures: 1-118. Lecture 1 "Delivered May '76" (This is not very distinct, the date is from the catalogue: probably the date of lecture 1 is Nov ii '76 making the session 1776-77).

The name Power is written on the verso of the last free end paper of Vol II. There is a bookplate with the monogram WJP.

52. **Glasgow: University of Glasgow, The Library**
    MS Ferguson 36   n.d.

"Lectures on Chemistry by Dr. Black" (verso of f 2)

One volume 8⁰ 18.5 x 11.0 cm; both sides, one hand; fly leaf + 22 blank leaves (numbered 1-23 in pencil), text starts on f 24 (numbered in pencil), ff [25-189], f 190 is a blank fly leaf.

Lectures: 5-19, not all of which are numbered. Lecture 18 mentions the experiments in Paris on the vaporization of diamonds (1773).

Bookplate of John Whitefoot Mackenzie. Acquired by Ferguson April 1886.

53. **Glasgow: University of Glasgow, The Library**
    MS Ferguson 40-41   n.d. (c1783)

No title

Two volumes 8⁰ 20.3 x 12.5 cm; one side, one hand; I: ff 1-196; II: ff 1-147.

Lectures: 1-39. Lecture 3 notes (f 23) "about the year 1781 the late Dr. Price . . ." (James Price died in 1783).

No name
Phillipps MSS 17837 on free end paper of each volume.

54. **Glasgow: University of Strathclyde,**
    Andersonian Library
                                            1767-68

"Notes From Dr. Black's Lectures on Chemistry 1767/8 Thos. Cochrane"
Two volumes 12⁰ 15.9 x 9.5 cm; one side, one hand; I: 652 pp; II: 688 pp.

Lectures: Not numbered or dated consistently; approximately 125.

Thos. Cochrane (attended 1767-68) was at the University from 1766 until 1769.
A transcription of this set of notes has been published privately: (Douglas McKie, editor) Thomas Cochrane *Notes from Dr. Black's Lectures on Chemistry 1767/8* (Imperial Chemical Industries Limited, Wilmslow, Cheshire 1966).

Refs: 6, 17 (entry NB 0520258: microfilm copy at Cornell University, Ithaca, N.Y.)

55. **St. Andrews: University of St. Andrews,**
    Chemistry Department Library
                                            1771-72

No title

Four volumes (Vols I, II, IV & VI) 4⁰ 22.9 x 17.8 cm; one side, one hand; I: 301 ff; II: 293 ff; IV: 308 ff; VI: 252 ff.

Lectures: 1-42, 61-78, 91-106. Lecture 1 dated Octr 29th 1771. The general introductory remarks are not included.

H.B.F. Beaufoy: Harry B.F. Beaufoy was enrolled 1771-72.

Refs: 2e, 5a, 20a , 25a, 26b

56. **St. Andrews: University of St. Andrews,**
    Chemistry Department Library
                                    n.d. (c1775)

"Chemistry Containing Whatever is Essential to be Known in the Art thereof Delivered in the University of Edinr By Dr. Joseph Black"

Two volumes 4° 23.0 x 17.7 cm; one side, several hands; I: 2 leaves (1st has notes by Dr. John Read, 2nd is the title), 321 ff; II: 260 ff, last a blank.

Lectures: 1-44. The dating is that of Dr. Read.

Bookplate of William Herbert (1718-1795), the bibliographer, over traces of a former plate in Vol I.

Refs: 2e, 5b, 20b, 24, 26a

GREAT BRITAIN, WALES

57. **Aberystwyth: University College of Wales,**
    The Library
    MS 50/B6   n.d. (1769-70?)

No title ("Black's Chemistry Vol. 5th" on the fly leaf)

One volume sm 4° 19.9 x 16.0 cm; both sides, one hand; pp 1-338 (146 is repeated, 262 left out, blank leaf after p 176, text not affected).

Lectures: 113-136. The date 1769-70 is suggested as a result of comparing lecture numbers and contents of this MS with No. 27.

No name

58. **Aberystwyth: University College of Wales,**
    The Library
    MS 50/B6   n.d.
    (boxed with No. 57)

No title

Four volumes (Vols I, III-V) 4° 23.0 x 18.5 cm; one side, several hands; I: 1 leaf index, pp 1-12, ff 13-189; III: ff 1-332; IV: ff 1-285; V: ff 1-291, with unnumbered single leaves after f 193 and f 253 and 4 blank leaves after f 199.

Lectures: 1-17, 18 (incomplete), 39-118. Lecture 64 has a reference to the translation of Scheele by Forster (1780).

J. Lyon: signature on first fixed end paper of each volume (a John Lyon was enrolled 1783-4, a James Lyon 1784-5 and a second (?) James Lyon 1784-5 & 1785-6).

UNITED STATES, CALIFORNIA

59. **Pacific Palisades: Private collection of**
    William A. Cole
                                    1771-72

"Lectures on Chymistry by Doctor Joseph Black Edinr 1772" (caption title p 1 Vol I)

Three volumes sm 4° 17.6 x 15.1 cm; both sides, one hand; I: pp 1-314 (i.e. 316 pp since numerals 27 and 314 are used twice); II: 1 leaf (notes on the Barometer, both sides), 2 blank leaves (2nd has an erased title), pp 2-307 (pp 82-85 misplaced after p 77, 108 used twice); III: title, pp 1-196.

Lectures: 1-124. Starts with the definiton of chemistry.

Angus Macdonald: signature and bookplate (enrolled 1771-72), and also the signature of W.S. Constable (Wm. in Vol II, W. in Vol III).

60. **Pacific Palisades: Private collection of**
    William A. Cole
                                    1788-89

"Lectures on Chemistry deliver'd at the College in Edinburgh by Dr. Black. Professor of Chemistry. 1788."

Three volumes (originally in 6, on the title: "Price for the 6 Volumes £5. s15. d6.") 4° 27.5 x 18.0 cm; one side, two or more hands; I: 5 leaves (1st is the title, 2nd Index (incomplete), last 3 leaves are blank), ff 1-229, 13 blank leaves, ff 1-108, [109-179]; II: ff [1-474]; III: ff [1-271], 10 blank leaves, ff [272-477].

Lectures: 1-118.

Bookplate of John Thomas Stanely of Alderley in each volume and his signature on the title. Stanley enrolled 1788-89, and was a friend of Dr. Black.

61. **Palo Alto: Stanford University,**
    Lane Medical Library
    MS C 28H/B62   1773-74

"Lectures on Chemistry delivered in the University of Edinb: by Dr. Black"

Eight volumes folio 31.3 x 20.0 cm; both sides, two (?) hands; I: pp 1-94; II: pp 95-184; III: pp 1-95; IV: pp 1-71; V: pp 1-96; VI: pp 97-192; VII: pp 1-96; VIII: pp 97-140. The inside of each cover is used for a table of contents for the volume.

Lectures: 1-118. Lecture 118 dated May 2 1774.

Initials LV (or SV) on covers of Vols I, II, VII & VIII.

62. **Palo Alto: Stanford University,**
    Lane Medical Library
    MS C 28H/B62/1777   1777-78

"A Course of Chemical Lectures as delivered by Dr. Black Professor of Chemistry in the University of Edinburgh *in 4 vols* 1777 *Vol 1*" (italicised words added in a different coloured ink)

Four volumes 8° 18.5 x 13 cm; both sides, one hand; I: pp 1-375; II: pp 1-381; III: pp 1-384; IV: pp 1-103; approximately 100 blank leaves.

Lectures: 1-59.

Signature of B.B. Brown in some volumes, also the signature of G.L. Simmons, Sacramento, and in Vol IV a rubber stamp: G.L. Simmons Medical Library.

63. **Redwood City: Private collection of**
    Dr. Roy G. Neville
                                    1777-78

"Lectures on Chemistry Read in Edinburgh by J. Black M.D. C.P. 1778 Vol: 3d"

One volume 8° 18.0 x 10.5 cm; one side, one hand; title, 288 ff (last a brief Contents).

Lectures: 42-64.

No name

Ref: 25c

UNITED STATES, CONNECTICUT

64. New Haven: Yale University,
               Beinecke Library
               MS 113   1768-69
               (Mellon Collection)

"A Course of Lectures on Chymistry delivered by Joseph Black M.D. Professor of Chymistry in the University of Edinburgh. Taken by Tucker Harris 1768/9 Vol. I."

Two volumes sm 4° 19.1 x 15.3 cm; both sides, one hand; I: 2 leaves (1st a note, 2nd title), pp 1-260; II: 261-514. At the end of Vol I, following a blank leaf, are 4 leaves Index Vol I.

Lectures: 1-127. Starts with the History of Chemistry. Lectures 1 and 2 dated Novr. 1st & 8th (?) 1768.

Tucker Harris (enrolled 1768-69, 1769-70, 1770-71).

Ref: 23a.

65. New Haven: Yale University,
               Beinecke Library
               MS 118   1773-74
               (Mellon Collection)

"A Course of Lectures on Chemistry delivered in the University of Edenburgh (sic) by Dr. Black. Oct: ye 28th 1773"

Four volumes 4° 23.3 x 17.9 cm; one side, one hand (except for Vol I prefatory matter); I: title, ff [i-v] Introduction, ff [vi-xxvii] lectures 1-4, ff 1-279; II: ff 280-612; III: ff 613-913; IV: ff 914-1168, 8 leaves Index (both sides, covers only ff 1-290).

Lectures: 1-118.

No name

Ref: 23b

66. New Haven: Yale University,
               Beinecke Library
               MS 125   n.d. (c1780)
               (Mellon Collection)

No title

One volume 4° 22.2 x 18.3 cm; partly one side, one hand; ff 1-174, pp 175-412.

Lectures: 1-39, not always clearly designated.

Bookplate (card) of W. Clement Daniel M.D., Epsom.

Ref: 23c

67. New Haven: Yale University,
               Beinecke Library
               Blagden Papers Box 7,   n.d.
               Osborn Collection

No title

Three "volumes" folio (in gatherings of 6 leaves, not bound) 32.2 x 20.5 cm; one side, three hands; I: ff 61-84, 2 leaves; II: ff 1-155; III: ff [1-280].

Lectures: Not numbered. Covers last part of Mixtures to Nickel.

No name
Owned by Sir Charles Blagden, and with what appear to be additions by him.

UNITED STATES, KANSAS

68. Lawrence: University of Kansas, Deprtment of
               Special Collections, Kenneth Spencer
               Research Library
               MS B10   1783-84

No title

One volume 4° (in 8's) 19.6 x 15.7 cm; both sides (except for occasional blank page or leaf at the end of a lecture), it appears to be in one hand; 255 unnumbered leaves.

Lectures: 57-89.

Peter Crompton: Edinburgh May 8 1784 (on a loose leaf, probably formerly the front free end paper). Crompton was enrolled 1781-1786.

UNITED STATES, MARYLAND

69. Bethesda: National Library of Medicine
               MS B/2   1773-74

No title

Two volumes 8° 19.5 x 12.6 cm; both sides, one hand; I: ff 1-355; II: ff 1-280; the leaves have been numbered by a machine.

Lectures: 1-118. Lecture 118 dated May 2 1774.

James Bruce Script in Vol II verso of f 280. This name also appears in No. 50 (1773-74). The name is on the rolls for 1772-73, 1773-74 and 1775-76 and also for the period 1788 to 1795. It has not been discovered how many persons are involved but it does seem likely that one of them was in the business of supplying copies of the lectures to other students.

Refs: 4, 17 (entry NB 0520251, which gives the date as Edinburgi 1795-96)

UNITED STATES, NEW YORK

70. Albany: The New York State Library
               MS V785/540.4/B62   1785-86

"Notes from Dr. Black's Lects. on Ch[emistry]" (caption title f 1, cropped)

One volume 8° 19.5 x 15 cm; generally one side, one hand; ff 1-301 (numbering is irregular, only 289 numbered leaves present) + 61 ff.

Lectures: 1-127. Lecture 1 dated October 26th 1785, last lecture April 18th 1786.

Received from the Utica State Hospital 1924. Signature of Dr. Brigham on the fly leaf. Bookplate: Medical Library of the New York State Lunatic Asylum No. B. 249.

Ref: 17 (entry NB 0520256)

71. Ithaca: Cornell University, Olin Library,
               History of Science Collections
               MS Lavoisier/QD/B62   1773-74

"Lectures on Chymistry, delivered by Joseph Black. M.D. Professor of Chymistry, in the University of Edinburgh. Vol. I."

Three volumes 4° 23.4 x 18.5 cm; one side, two hands; I: title, blank leaf, ff 1-411, 2 leaves Index; II: title, blank leaf, ff 1-454; III: title, blank leaf, ff 1-412 + laid in leaf outline of index.

Lectures: 1-118. Lecture 118 dated 2 May 1774.

Joseph Fryer [of] Rastrick on the title, also Joseph Fryer on the end paper of Vol II. Fryer enrolled 1776-77.
This is the copy once owned by Professor R.C. Gale and sold by him to Henry Sotheran and Co., London (Catalogue 894 (1951) item 1101). The manuscript was purchased by Mr. Denis I. Duveen and later sold to Cornell University. Mr. Duveen gave it the date 1770 with the owner's name as James Freyer Bostwick.

Refs: 5c, 16

72. **Ithaca: Cornell University, Olin Library, History of Science Collections**
MS Hist Sci/QD/39/B62   n.d. (c1780)

"Notes from Dr. Black's Lectures on Chemistry. By Jacob Pattison"

Two volumes 8° 18.8 x 11.5 cm; one side, one hand; I: title, 137 ff, several are blank; II: 82 ff (+ blank leaves and notes on magnetism and electricity + (inverted) medical notes).

Lectures: Only a few are numbered.

Jacob H. Pattison LLB LLM Cantab: Pattison enrolled at Edinburgh 1779-80.

73. **New York: The New York Academy of Medicine**
n.d. (c1780)

"Lectures on Chemistry, delivered in the University of Edinburgh"
(a name and date have been cut from the title)

Three volumes 4° 22.5 x 17.7 cm; both sides, one (?) hand; I: pp 1-512; II: pp 1-516; III: pp 1-476 (+ 38 pp of Dr. Monro's New system of the nerves Lect: 13 Jany 1779). Each volume has a title before p 1.

Lectures: 1-118. Tentative date of the manuscript is from the remains of the date on the title and from the date of Alexander Monro's lecture.

No name

UNITED STATES, PENNSYLVANIA

74. **Philadelphia: College of Physicians of Philadelphia**
MS 10a/12   n.d. (c1785)

No title

Four volumes 8° 18.3 x 11.0 cm; both sides, one hand; I: pp 1-400; II: pp 1-396; III: pp 1-400; IV: pp 1-421.

Lectures: 1-118.

Caspar Wistar (1760-1818) graduated M.D. University of Edinburgh in 1786 and became professor of chemistry at Pennsylvania College in 1789. Presented to the College of Physicians of Philadelphia in the name of Dr. Mifflin Wistar, 1889.

Ref: 17 (entry NB 0520260, which gives the date as 1784)

75. **Philadelphia: College of Physicians of Philadelphia**
MS 10a /13   1792-93

"Syllabus of a Course &c of Chemistry by Dr. Black"

One volume 8° 18 x 11 cm; one side for the most part, one hand; ff 1-81.

Lectures: Unnumbered, but dated from Nov. 12, 1792 to Jan. 26, 1793 (ff 1-56 "Definition of Chemistry" to "Expt on Magnesium & Quick Lime"). ff 57-81: undated lectures on properties of various minerals.

No name
This MS is possibly in the same hand as No. 76.

76. **Philadelphia: College of Physicians of Philadelphia**
MS 10a/14   1792-93

"Notes from Dr. Black's Lectures on Chemistry"

One volume 8° 19.5 x 12.2 cm; one side, one hand; 49 ff + blank leaves (except for medical notes with the book inverted).

Lectures: Not numbered. Starts with the composition of spirit of wine and ends with lead; about 29 lectures in all. The first lecture is dated February 27th.

Tho. C. James (in pencil on the title page): Thomas Chalkley James (1766-1835) was enrolled 1792-93. Bookplate of Hugh Lenox Hodge M.D. (1796-1873). Acquired by the College of Physicians May 15, 1896.

Ref: 17 (entry NB 0520257, dated at Phila. (sic) 1793).

77. **Philadelphia: The Historical Society of Pennsylvania**
MS Yi 2/1603 Q   1773-74

No title

Three volumes 4° 23.0 x 18.0 cm; both sides, one (?) hand; I: pp 1-395; II: pp 1-403; III: pp 1-485.

Lectures: 1-118. Lecture 29 dated 15th Decmr 1773.

William Logan. Gift of George Logan d. April 9, 1821.

Ref: 17 (entry NB 0520229, gives date as about 1770 and location as the Library Company, Philadelphia: all MS materials have however been sent from there to the Historical Society)

78. **Philadelphia: The Historical Society of Pennsylvania**
MS Am/3313   1786-87 ?

"Lectures on Chymistry by Joseph Black M.D. Professor of Medicine in the University of Edinburgh"

Two volumes 4° 24.3 x 17.2 cm; partly on both sides, one hand; I: 225 ff, 1 blank leaf, pp 1-12, 1-22, 1-40 (all are texts of lectures); II: 163 ff + lectures on Philosophy of Chemistry and Natural History by Henry Noyes Phila. Feb 18 1785.

Lectures: Not divided into lectures.

Wm Martin 1785 on title Vol I. William Martin (1765-98) was probably the William Martin enrolled in 1786-87 and 1787-88.

Refs: 3a, 14, 15

79. **Philadelphia: Pennsylvania Hospital Historic Library**
MS QD 39/B56a   1771-72

"Lectures on Chemistry By Joseph Black MD Professor of Chemistry In The University of Edinburgh 1771 Vol 1st"

Two volumes 4⁰ 23.5 x 19.0 cm; both sides, one hand; I: 1 leaf (title, verso a note), pp 1-600, 4 blank leaves; II: 2 blank leaves, 1 leaf (title, verso blank), pp 1-539 (actually 542 since numbers 130,458 and 478 are repeated), 11 blank leaves.

Lectures: 1-131.

Thomas Parke (enrolled 1771-72): given to the Pennsylvania Hospital by Thomas Parke.

Verso of title Vol I: "Note. These lectures were delivered in Edinburgh by Dr. Black taken down in shorthand by a Society of Students who daily compared their Notes, & are upon nice inspection found a very correct Copy of that great and accurate Chemists Lectures. Edinburgh May 1772 Thomas Parke"
At the end of Vol I (lecture 80): "Finis Vol 1st Edinburgh Jan 7, 1772 taken in shorthand by T. Parke esq" and at the end of Vol II: "Finis Edinburgh May 1772 taken in shorthand by T. Parke & Co."

80. **Philadelphia: University of Pennsylvania Library**
MS 14, Edgar Fahs   1775-76
Smith Memorial Collection

No title

Six volumes 4⁰ 22.3 x 18.5 cm; one side, one (?) hand; I: ff 1-11, [12-260]; II: ff [1-268]; III: ff [1-255]; IV: ff [1-264]; V: ff [1-244]; VI: ff 1-251.

Lectures: 1-118. Lecture 29 dated Dec. 15 1775.

Presented to the Society of Apothecaries by Jac. Hill Esq. 1832.

UNITED STATES, WISCONSIN

81. **Madison: University of Wisconsin, Memorial Library, Rare Books Department**
MS M 11a   1776-77

"Chemistry by Dr. Black 1776 by Dr. J. Clarke" (verso f 119)

One volume 8⁰ 19.1 x 12.0 cm; both sides, one hand; ff 1-119 (+ other notes on medicine, including Dr. Homes Lects on Materia Medica Oct 31 '78, etc).

Lectures: Approximately 118, mostly unnumbered. Lecture "71" on verso f 50 dated Feby 28th 77, a few others are dated also.

Joseph Clarke (1758-1834) studied in Edinburgh 1776-79, graduated M.D. 1779, and became a celebrated obstetrician.

Refs: 3b, 17 (entry NB 0520250)

UNITED STATES, TEXAS

82. **Galveston: University of Texas Medical Division**
    1773-74

"Dr. Black's Lectures on Chemistry A.D. 1774"

Two volumes 8⁰ 20 x 16 cm; both sides, two hands; I: title, pp 1-348; II: title, pp 349-429, [430-486], 1 blank leaf, pp 1-208.

Lectures: 1-100; not all are numbered.

No name

AUSTRALIA

83. **Victoria: Private collection of Air Commodore F.J.P. Wood (retired)**
    n.d.

"Lectures on Chymistry by Joseph Black MD"

One volume folio 35 x 26 cm; both sides, one hand; title, pp 1-579, 84 ff, pp 1-23 (drawings of apparatus and tables) comprising Appendix.

Lectures: Not numbered. Starts with Heat. On p 579 is the start of the section on Antimony but it breaks off in mid-sentence. A note follows: "I am to go on from this place till the acct. of Metals is finished which will complete this Copy – The Water & Vegetables being finished already." The 84 un-numbered pages cover Water etc, ending with the statement: "The learned and very ingenious Doctor now concluded the Course with some Remarks on the more general Principles of the art of Pharmacy, but which, as they contain nothing new or un-common, were purposely left untranscribed."

Wm Wood: signature on title. The Appendix has a 'title' leaf with "Joseph Black MD Edinburgh Thomas Jones R. Reece". Some of the drawings have page references to Reece: this work has not been identified as yet. Photocopy at University of Melbourne.

CANADA

84. **Toronto: University of Toronto, Thomas Fisher Rare Book Library**
MS 3192   n.d. (c1780)

"Lectures on Chemistry By Joseph Black M.D. And Professor of Chemistry in the University of Edinburgh Volume 3"

One volume 4⁰ 22.5 x 18.5 cm; one side, two hands; total of 344 ff.

Lectures: 57-87. Lecture 57 continues the topic Magnesia Alba. An addition reads: "in the year 1780 when making some experiments . . ."

No name. Jos. Black written on the last leaf is not in Dr. Black's hand.

ISRAEL

85. **Jerusalem: Jewish National and University Library**
The Sidney M. Edelstein   n.d. (c1785)
Collection

No title

Six volumes 8° 18 x 11 cm; both sides, two hands; I: 517 pp; II: 548 pp; III: 528 pp; IV: 572 pp; V: 402 pp: VI: 449 pp. Laid in in Vol I is an Index in Howcutt's hand.

Lectures: 1-118.

John Howcutt Buffalo, N.Y. and Denver Colorado 1883.
"These books belonged to my friend Professor Geo. Hadley."

Ref: 21. The last bookseller to have the MS dated it c1785 based on the York Medical Society MS (see list of unlocated manuscripts below). The date 1766 is too early and is apparently based on a note stating that these are lectures given by Dr. Black "commencing in the year 1766". In the text (p 26 Vol III) is: "Mr. Schele explained the nature of tartar": this was in 1770. Other parts of the MS have not been consulted.

NEW ZEALAND

86. **Dunedin: Library of the University of Otago Medical School**
1791?

"Chemical Lectures by Dr. Black of Edinburgh"

One volume folio 32.7 x ? cm; both sides (?), one hand (?); title, (inscription on verso), first 6 numbered pages torn out, pp 7-609, 1 blank leaf, 3 pp Index (233 missed out, p 548 is blank).

Lectures: Called a "complete course".

Ragner Bellman Aug. 9 1791: at the bottom of p 609. There is no record of him as a student at the University of Edinburgh.

Ref: 18

87. **Dunedin: Library of the University of Otago Medical School**
MS Monro Collection Vol 245   1796

"Notes from Doctor Black's Chymistry, Vol II 1796"

One volume 4° 23.5 x ? cm; both sides (?), one hand; first leaf torn out, second leaf Index (both sides), third leaf title (on verso), fourth leaf torn out, pp 1-322, many leaves cut out at the end.

Lectures: Covers Metals and some topics in Vegetable chemistry.

Alexander Monro *tertius* (1773-1859), M.D. Edinburgh 1797.

Ref: 18

ADDENDA

GREAT BRITAIN, SCOTLAND

32a. **Edinburgh: National Library of Scotland**
Acc 4796.113   1775

"Notes of Lectures on Chemistry delivered by Dr. Black at Edinburgh, begun 1st March 1775"

Two volumes 4°; I (collection of 4 page gatherings of 2 sheets each, prepared for binding): 21.0 x 16.5 cm; one side, one hand; ff 1-333 (lacks ff 198-201); II (vertical format shorthand notebook with interleaved blotting paper): 9.5 x 17.0 cm; both sides, same hand; 78 ff (ff 70-72 are stubs).

Lectures: 1-34 (Vol I), 35-40 (Vol II); with the notebook reversed the stub of f 72 is marked 'Lecture 6th'.

Sir William Forbes, 6th Bt. of Pitsligo (1739-1806). Acquired by the National Library of Scotland in 1969.

36a. **Edinburgh: Royal College of Surgeons of Edinburgh**

1770

"Lectures on Chemistry Delivered by Dr. Joseph Black Professor of Chimistry in the University of Edinburgh 1770 VOL:I:"

Two volumes 4°; I: 21.5 x 18.0 cm; both sides, one hand (but with the title written in the hand of Vol II); pp 1-477, 7 blank leaves; II: 24.0 x 19.5 cm; both sides, second hand; pp 1-237, 4 blank leaves.

Lectures: Not numbered. Vol I ends with Earths; Vol II continues from Inflamable Bodies to Animal Substances.

No name
Listed in *Catalogue of the Library of the Royal College of Surgeons of Edinburgh* (Edinburgh, 1863).

APPENDIX 1: UNLOCATED MANUSCRIPTS

1. CAYLEY, George (1763-1831): Six volumes 1311 total leaves, 118 lectures, dated 1785-86. Cayley attended Black's lectures in 1785-86 and graduated M.D. 1789. The last location was the York Medical Society Library, but they were unable to locate the manuscript in 1974 or again in May of 1979. This manuscript was used extensively by Dr. Douglas McKie in his Black manuscript studies in *Annals of Science*.

2. RAIT, Will: One volume 102 pp (with four sets of notes on medical topics) covering approximately lectures 1-14, and dated 1773-74. William Rait was enrolled 1773-74. This manuscript was last seen in the possession of a bookseller in Los Angeles.

3. RAMSAY, William: See his *Life and Letters of Joseph Black, M.D.* (London 1918) p 12: "I myself possess a copy of manuscript notes of "Lectures read in Edinburgh by Joseph Black, M.D. in 1773", It is well bound and written by an unknown scribe in an excellent hand." This manuscript was not with the Ramsay library when Henry Sotheran & Co. acquired and sold it circa 1925.

4. ROBISON, John: In a letter to George Black Jr., Jan 20, 1800, Robison refers to "a Copy of Notes taken by some Student in 1773, bound in four Volumes, . . .". This manuscript was with Black's papers. See Eric Robinson and Douglas McKie *Partners in Science. James Watt & Joseph Black* (London 1970) p 333. Also see Joseph Black *Lectures on the Elements of Chemistry* (London and Edinburgh 1803) Vol I p xi, where Robison states: "I had the assistance of a very fair copy of notes, taken by a student, or rather manufactured by the comparison of many such notes. Copies of this kind were to be purchased for four or five guineas. This copy belonged to Dr. Black, and he had made many alterations and insertions of whole pages with his own hand."

5. ROBISON, John: in a footnote on p 398 of Vol II of his edition of the printed lectures, Edinburgh 1803, he cites Black's thoughts on the increase of weight on calcination from a Glasgow lecture manuscript of 1762.

6. THOMSON, Thomas (1773-1852): See 'Biographical Notice of the late Thomas Thomson' *Glasgow Med. J.* 5 pp 69-80, 121-153 (1858), p 75, and J.R. Partington *History of Chemistry* Vol III (London 1962) p 716. Thomson is said to have left three 8° volumes of notes. He attended Black's lectures in 1795-96.

7. "An Abstract of Doctr. Black's Lectures on Chemistry" c1770 one volume 8° 18 leaves, a resume of 10 lectures — of the more general effects of Heat, defines chemistry as the "study of the effects of heat and mixture upon all bodies, whether natural or artificial": see H. Zeitlinger [*Sotheran's*] *Bibliotheca Chemico-Mathematica* (London 1921) p 22, item 426.

8. HALLIWELL, James O. *A Catalogue of Scientific Manuscripts in the possession of J.O. Halliwell, Esq.* (London 1839), items 4-6 Notes of Dr. Black's Lectures on Chemistry three volumes 8° n.d. See: A.N.L. Munby *The History and Bibliography of Science in England: The first phase, 1833-1845* (Berkeley, California 1968) p 33.

APPENDIX 2: DATES OF THE MANUSCRIPTS IN THIS SURVEY

Manuscripts having uncertain dates are entered in italics. The superscript (*) denotes "incomplete", meaning 85% or less of the lectures are present. 16 of these have 25% or less and 16 are 26-50% complete.

1766-67: 5*, 19	1780-81: *66*, 72*, 73, 84*
1767-68: 30, 54	1781-82: *2*, 41*
1768-69: 22, 64	1782-83: 42, 43*
1769-70: 9, *23*, 27, 31*, 57*	1783-84: 44*, *53*, 68*
1770-71: 3, 36a, 37	1784-85: —
1771-72: 38*, 55*, 59, 79	1785-86: 35, 70, *74, 85*
1772-73: 7*	1786-87: 6*, *47*, 78*
1773-74: 11*, *12*, 13, 16*	1787-88: *45*
17*, 18, 50, 61, 65,	1788-89: *36, 49*, 60*
69, 71, 77, 82*	1789-90: —
1774-75: 24, 39	1790-91: 4
1775-76: 1*, 14, 32a*, 40, *56*	1791-92: *86*
80	1792-93: 75*, 76
1776-77: 20*, *51*, 81	1793-94: —
1777-78: 21, 33*, 62*, 63*	1794-95: —
1778-79: 15	1795-96: 25
1779-80: —	1796-97: 46, 87*

The following have not been ascribed to particular sessions:

8, 10, 26*, 28, 29*, 32*, 34, 48*, 52*, 58*, 67*, 83*